ISLANDS in INFINITY

Galaxies 3-D

THE
London Stereoscopic Company, LTD.

Published in 2026 by
The London Stereoscopic Company (LSC)

A catalogue record for this book is available from the British Library.
UK ISBN 978-1-8381645-6-0

Published in North America by
University of Arizona Press (UAP)
Main Library Building, 5th Floor
1510 E. University Blvd, Tucson, AZ 85721-0055
www.uapress.arizona.edu

A Library of Congress control number is available.
ISBN 978-0-8165-5630-4 (Hardcover) and
978-0-8165-5631-1 (Ebook)

The London Stereoscopic Company
Director – Brian May
Publisher and Editor-in-Chief – Robin Rees
Art Director and Illustrator – James Symonds
Astrophysics Consultants – Prof. Victor P. Debattista and Prof. Steven N. Longmore
Cosmology Consultant – Dr Mark Norris
Editor – Bill Hemsley
Proofreading – Florence Production Ltd
Astronomy Proofreader – Glenn Smith
LSC Manager – Sara Bricusse
PR Manager – Nicole Ettinger
PA to Sir Brian – Sally Avery-Frost
Website – Phil Murray

Book printed and bound in Wales by Gomer Press.
OWL Stereoscope designed by Brian May.
First printing 2026.

www.LondonStereo.com
www.BrianMay.com
BrianMayForReal
@DrBrianMay
@LondonStereo
@LondonStereo

Note about the images

Unless otherwise indicated all the 3-D photographs in this book are by J.-P. Metsävainio.
All illustrations are by James Symonds unless otherwise credited.
Monos are individually credited.
Every effort has been made to correctly credit the photographers who have contributed to this work. If there are any wrong attributions, please contact the publishers.

This book is dedicated to Sir Patrick Alfred Caldwell-Moore, the 'grandfather' of astronomy education and inspiration to generations of astronomers, ourselves included.

CONTENTS

Preface

During my 2017 visit to *Astrofest*, the annual London-based astronomy festival, on display on the London Stereoscopic Company's (LSC) stand I viewed some stereo cards of nebulae, the beautiful clouds of gas and dust that hang in the night sky, like pictures at an exhibition, and give birth to new stars. I was blown away by the sheer beauty of these objects as they sprang into 3-D in Sir Brian May's stereoscopic viewer. Space is of course three-dimensional, so for the first time I felt I was seeing them as they actually appear. A seed was sown! This was a new way forward for sharing the joy of astronomy with the world at large.

In 2020, a beautiful book of images of the nebulae was published by the LSC, written by David Eicher and Brian May, with extraordinary 3-D images of nebulae captured and created by the artist and astrophotographer J.-P. Metsävainio. *Cosmic Clouds 3-D* set a new standard for astrophotography, and it was not long before we began thinking about a similar title regarding galaxies. Not only would the images be spectacular, but the book would, as Brian had said, 'leak' into explaining cosmology, the history of the Universe from the Big Bang to the present day. This book would therefore not only tell the story of galaxies, but also be the one-stop guide to understanding the Universe.

In 2023 we began developing the book and, following a period of detailed work with Brian, J.-P. Metsävainio and technical artist James Symonds, we present *Islands in Infinity: Galaxies 3-D*. It incorporates images garnered from many different Earth- and space-based observatories, some requisitioned especially for the task of photographing specific southern hemisphere galaxies needed to present a complete picture of the diversity and nature of these *Islands in Infinity*.

In this book you will, I hope, find everything you need to understand what galaxies are, how they originated, how they have evolved into many different types of galaxy, and how they interact and give birth to everything we experience on our tiny blue planet inside our Solar System, itself inside our very own Galaxy, the Milky Way.

Peering out, we see independent galaxies outside our Milky Way, and looking still further afield we see how they clump together in clusters, superclusters of clusters, and great walls of galactic filaments and hot plasma as we head out in space and back in time towards the origins of space and time themselves at the Big Bang.

I hope you will enjoy this journey. It is the greatest story of all, and the more we understand our origins and how infinitesimally small are our human concerns on a cosmic scale, the better our chances of enjoying our time on Earth with humility and wisdom.

Derek Ward-Thompson

The Milky Way and the Atacama Large Millimeter/submillimeter Array (ALMA)

This view shows several of the ALMA antennae, located in northern Chile, with the central regions of the Milky Way above. The image was captured during the ESO Ultra High Definition (UHD) Expedition.

Introduction

The London Stereoscopic Company (LSC) is proud to offer this publication, *Islands in Infinity: Galaxies 3-D*, which we believe to be the very first attempt in the history of the Universe to offer a comprehensive state-of-the-art study of what we currently know about galaxies illustrated with stereoscopic three-dimensional (3-D) images.

Why has this not been attempted before? Because until now it has not been thought possible.

Those of you who have followed the activities of the LSC know that our mission is to propagate knowledge and beauty, science and art in 3-D. In the world of astronomy we have already offered the Apollo missions (*Mission Moon 3-D*), the OSIRIS-REx mission to asteroid Bennu (*Anatomy of an Asteroid*) and a study of nebulae (*Cosmic Clouds 3-D*). Pushing ever further outwards into the known Universe, we set our sights on a stereoscopic survey of those vast agglomerations of hundreds of billions of stars once known as 'island universes' and now called galaxies.

Following on from *Cosmic Clouds 3-D*, our strategy in producing this study of galaxies has been to bring together two leaders in their fields. In this case we have brought together to write the text an expert in the study of galaxies for over 40 years, Professor Derek Ward-Thompson, and for the illustrations, astrophotographic grandmaster J.-P. Metsävainio – the same J.-P. who produced the extraordinary stereos for *Cosmic Clouds 3-D*.

At an early meeting with Derek, we were talking about how to visualize the huge distances we are always faced with in astronomy. He presented the following analogy: if our Sun were shrunk to the size of an orange in London, then to the same scale the closest star would be another orange... in Moscow.

But what of galaxies? He then suggested we imagine our Milky Way Galaxy as a dinner plate on a dining table. Where would our nearest neighbouring galaxies be? Well, surprisingly, to us at least, only a few feet away – indeed there might be many galaxies around the table. So in an instant we had a picture in our minds of how these whirlpools and diffuse clouds of stars fill the whole Universe, relatively closely packed, as far as we can see and as far back in time as we can look. If we can understand how galaxies work, we can understand how the entire Universe works.

So, returning to the difficult problem of photographing galaxies in 3-D, of course there is no way to actually capture images from the two viewpoints separated by the baseline of thousands of light-years that would be required to make a natural stereo image. But our unique astro-stereoscopist J.-P. Metsävainio has found another way. He has spent years perfecting his technique of generating parallax shifts through the application of known distance data and superimposing them on a mono image to make its stereo partner. The results are breathtaking, and can be experienced first-hand by the reader using the OWL stereoscope supplied with this book.

For *Islands in Infinity*, J.-P. has created beautiful stereo images of every type of galaxy, and for the first time we can see galaxies as they actually are, in three dimensions. The images are scientifically correct as well as being aesthetically beautiful. They make it instinctively far easier to understand the structures, origins and evolution of galaxies, and how they merge to form beautiful cosmic artworks, such as the Porpoise and the Antennae. Every type of galaxy is described and shown here: spirals, barred spirals, ellipticals, irregulars and exotic galaxies.

To create his extraordinary stereos, J.-P. Metsävainio has photographed the night sky from his private observatory for around 30 years. He processes his images using painstaking manual editing and highly sophisticated software to create the stereo effect. He explains his methods in a section at the back of the book: 'How I Made the Stereos'.

For context, the book begins with an outline of fundamental astronomy and introduces some historical pioneers of the subject. It describes the cosmology of the Universe as we currently understand it, starting at the Big Bang and continuing through its evolution to the limits of current understanding.

Our LSC mantra remains: 'No Home without a Stereoscope'. So now please extract your OWL 3-D viewer from the backboard of the book and begin your in-depth exploration of the Universe!

Brian May

Sir Brian May looking into the eyepiece of the 28-inch Great Equatorial Telescope at the Royal Greenwich Observatory, London.

Galaxies M81 and M82

M81 is seen face on, with M82 below and edge on. M81 spans 100,000 light-years. The pair are gravitationally locked together and in the next few billion years will merge. Galaxy NGC 3077 can be seen slightly above and to the right of the large spiral, and NGC 2976 at upper left in the frame is face on. The field is seen through foreground nebulae. Total exposure time was around 40 hours.

Our Place in Space

1

Cataloguing the Night Sky

Light Pollution

Look up on a clear, dark night and you may see the Moon, depending on its phase, possibly a planet or two, and some bright stars. Even the most distant of these objects are very close by, in astronomical terms, at distances of a few to perhaps a few tens of light-years (the distance light travels in one year). For most people today that is the total of what is visible, astronomically speaking, with many rarely seeing even that much.

Among the many forms of pollution that humanity has inflicted upon planet Earth, light pollution is now one of the most obstructive to astronomical research and the appreciation of our place in the Universe. Very few people in the Western world have the luxury of growing up under dark skies, and the problem is getting worse.

However, if you are fortunate enough to live or visit somewhere with a reasonably dark sky, then at the right time of year you will be rewarded with a view of the band across the sky known as the Milky Way. This is the sum of light from large numbers of stars right across the astronomical body that we call our Galaxy. This word comes from the Greek *'galaxias'* meaning 'milky', and thus both terms refer to the milky-white appearance of this band of stars.

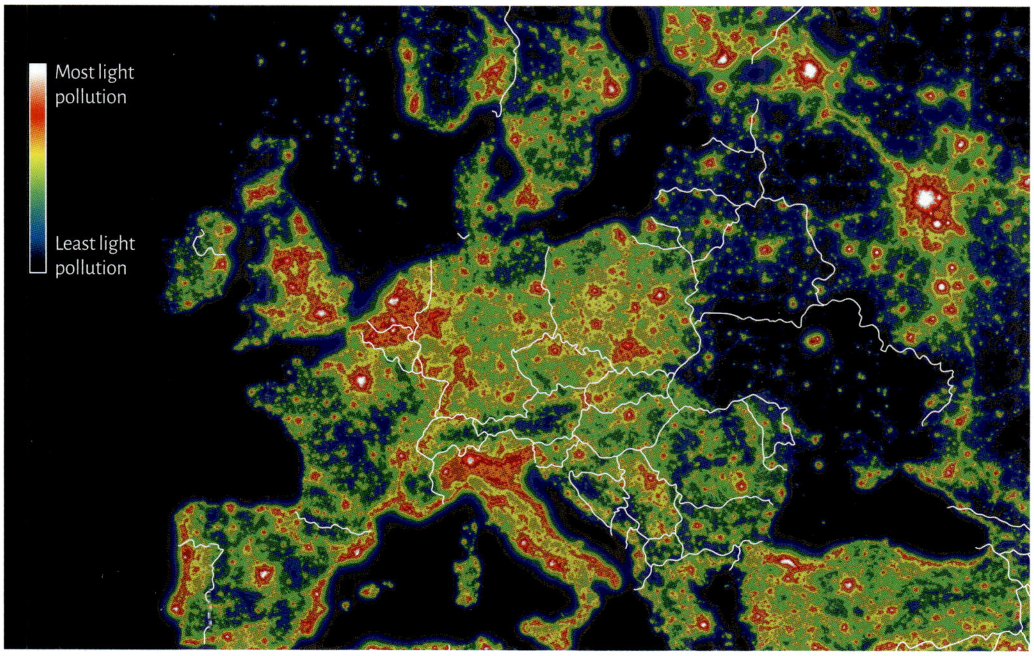

A light-pollution map of Western Europe, showing how few areas still have dark skies (the areas in dark blue).

The Capitalization of 'Galaxy'
The debate whether or not to use a capital letter is normally resolved by using the capital letter 'G' for Galaxy only when referring to our own Milky Way Galaxy, and to use the lower case 'g' for galaxy when referring to all other galaxies. We will follow this convention.

Cepheid Variable Stars
A Cepheid variable star is a type of star that has a well-defined relationship between its true brightness and the period of the variability in its brightness. This allows its distance to be estimated by comparing how bright it appears from Earth.

Billion and Trillion
We will use the modern definition of a billion, meaning one thousand million. We will also use the word trillion to mean one thousand billion. This will help us to avoid confusion when discussing distances and timescales in astronomy.

Nebulae or Galaxies?

When it was realized in the early 20th century that our Milky Way Galaxy is one of many, many similar objects, the term 'galaxy' was used to describe them all. The fact that this realization came as late as the 20th century may be a surprise to some readers, but before that time there was a debate as to whether these 'nebulae', as they were then known, were contained within our Galaxy or outside it. The discussion continued into the 1920s, with Heber Curtis and Harlow Shapley notably debating whether the Andromeda galaxy, or nebula as it was then known, lay within our Galaxy or not. This was resolved in 1923 when Edwin Hubble discovered Cepheid variable stars (see box above) in the Andromeda nebula/galaxy and thereby proved that it was further away than the far side of our Galaxy.

The Limits of the Human Eye

Returning to our observer standing under a sufficiently dark sky, if their eyesight is sharp enough, they may be able to see the Andromeda galaxy. It is certainly visible in the right conditions through a small telescope or a decent pair of binoculars. It appears only as a faint, fuzzy patch. Nevertheless, this is the furthest object that the human eye can see unaided, at a distance of 2.5 million light-years.

Galileo Galilei (1564–1642), the inventor of the astronomical telescope, shows the doge of Venice how to use his instrument. Painting by Giuseppe Bertini (1825–98).

An observer in the southern hemisphere at a dark site can see two additional galaxies with the unaided eye: the Large Magellanic Cloud (LMC) and the Small Magellanic Cloud (SMC). Named after the 16th-century explorer Ferdinand Magellan, these two apparent clouds are now known to be galaxies just beyond the Milky Way, described as 'satellite galaxies' of our Galaxy. They are somewhat irregular in shape, earning the SMC the classification of 'irregular', although the LMC is now classified as a spiral galaxy (we discuss galaxy classification later).

All other galaxies require a telescope to observe them, so the study of galaxies can be said to belong to the 'telescopic era', following the introduction of the astronomical telescope by Galileo just over 400 years ago. In fact all but the closest galaxies are revealed only by photographic exposures of many hours using the biggest modern telescopes. In these remarkable 'deep field' photographs, galaxies are revealed as stunningly beautiful objects – glorious islands in infinity.

Astronomers estimate that there are about 100 billion stars in our Milky Way Galaxy, which measures about 100 thousand light-years across. And now it is estimated that there may be 100 to 200 billion galaxies in the observable Universe, in which the furthest galaxy is seen as it was over 13 billion years ago, only a few hundred million years after the Big Bang, when space, time and the Universe as we know them began.

Lord Rosse's great reflecting telescope at Birr in Parsonstown, Ireland, with its giant 72-inch primary mirror. M51, the Whirlpool Galaxy, appears, sketched by Rosse himself, in the top left-hand corner of the upper image. His sketch revealed the spiral nature of the 'nebula', as it was then known, and that it was made up of individual stars. Below is a contemporary stereo card of the famous instrument, photographed by Lady Mary Rosse.

Stereoscopy

In this book you will see galaxies as you have never seen them before. We will travel from the birth of galaxies through the history of the Universe to the array of galaxies we see today. We will see galaxies colliding. We will explore many different kinds of galaxies and their evolution. We will look at galaxies near and far, galaxies young and old, and enjoy the beauty and majesty of the Universe.

But there is more! In this book you will for the first time see these objects in three dimensions, through the magic of stereoscopy, otherwise known as '3-D'. Our method of achieving this, which is basically Victorian stereoscopy, uses two adjacent flat images on a page that, with the help of a simple stereoscopic viewer, communicate to us the same experience of distance, depth and solidity that we enjoy all through our lives when perceiving our immediate surroundings. It works by separately presenting two slightly different images, one to each eye, in which the object is seen from slightly different points of view. So, just as in real life, our brains combine the two images and instantly decipher the slight 'parallax' differences to build a three-dimensional image in our heads.

Charles Wheatstone (1802–75) discovered stereoscopy and made these drawings for the stereoscope in 1838, before photography had become established.

Taking stereo pairs of images here on Earth is relatively easy: we just take two photos from positions about 2.5 inches apart – moving the camera sideways about 2.5 inches – the distance between our eyes.

In space, however, stars and galaxies are so remote that we cannot move the camera far enough sideways between shots to change the angle of view significantly. So J.-P. Metsävainio has evolved a technique that, starting from one captured 'flat' photograph, creates a second image to complete the stereo pair. It is done using available data and special software to impose a small displacement on every piece of the image. The result is truly magical! This might be a good moment to retrieve your OWL stereoscope from the back of this book. Use it to view any pair of images (those with rounded corners) in these pages and enter 3-D space! If you need help, there are some instructions on the page that houses the viewer.

Distances in Space

Astronomers use a variety of measures of distances. In everyday situations, miles, metres or kilometres can be used. However, in astronomy the numbers involved very rapidly become so large as to be practically unusable. So astronomers have invented larger units of distance. The first is the astronomical unit (AU), the distance from Earth to the Sun. Then there is the light-year, or the distance travelled by light in a year. Astronomers also use parsecs, which are the result of using Earth's motion around the Sun to triangulate to the distances to the nearest stars – see the explanatory box below. In this book, we will mostly use light-years.

Further Away is Further Back in Time

As we look to more and more distant galaxies, the time taken for the light to get to us from them increases. So we see them as they were in the past. And so, the further away we look, the further back in time we see.

Astronomical Units (AU)

Astronomers refer to units known as astronomical units (AU), where 1 AU is simply the average distance from Earth to the Sun, or about 150,000,000 kilometres.

Light-Years

A light-year is the distance that light travels in one year at a speed of about 300,000 kilometres per second, or about 186,000 miles per second. This enormous speed means that light travels just under 10 trillion kilometres (just under 6 trillion miles) in a year. This distance is what we call a light-year. Closer to home, a light-minute is the distance travelled by light in 1 minute. The Sun is about 8.3 light-minutes from Earth.

Parsecs

As Earth orbits the Sun once every year, it travels from a position 1 AU on one side of the Sun to 1 AU on the other side of the Sun and back again. This can be used to triangulate on the nearest stars, the way that surveyors triangulate to map areas of Earth. In this case the baseline of the triangulation is the radius of Earth's orbit, so the nearest stars appear in a slightly different place as we move from one side of Earth's orbit to the other. This is known as 'parallax'. Typically the movement is tiny and is measured in seconds of arc, where 1 second is 1/3,600 of a degree. Nevertheless, it can be measured. If a star shows a parallax of 1 second of arc, it is said to be at a distance of 1 parsec. It transpires that 1 parsec is just over 3 light-years.

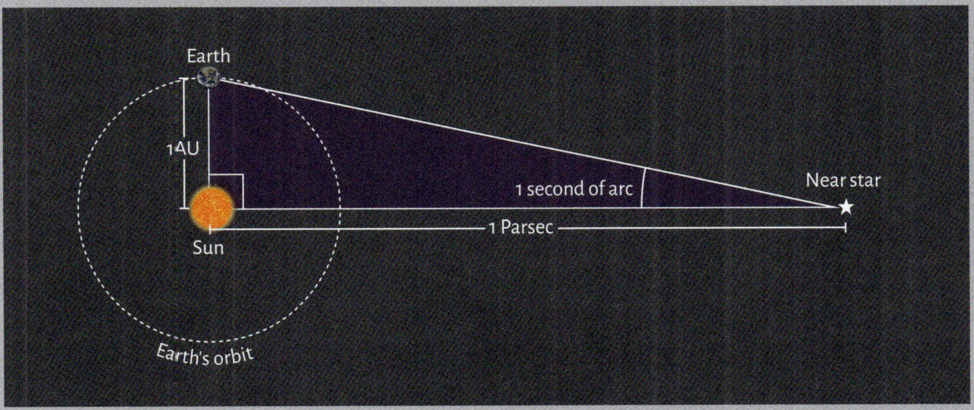

The length of 1 parsec.

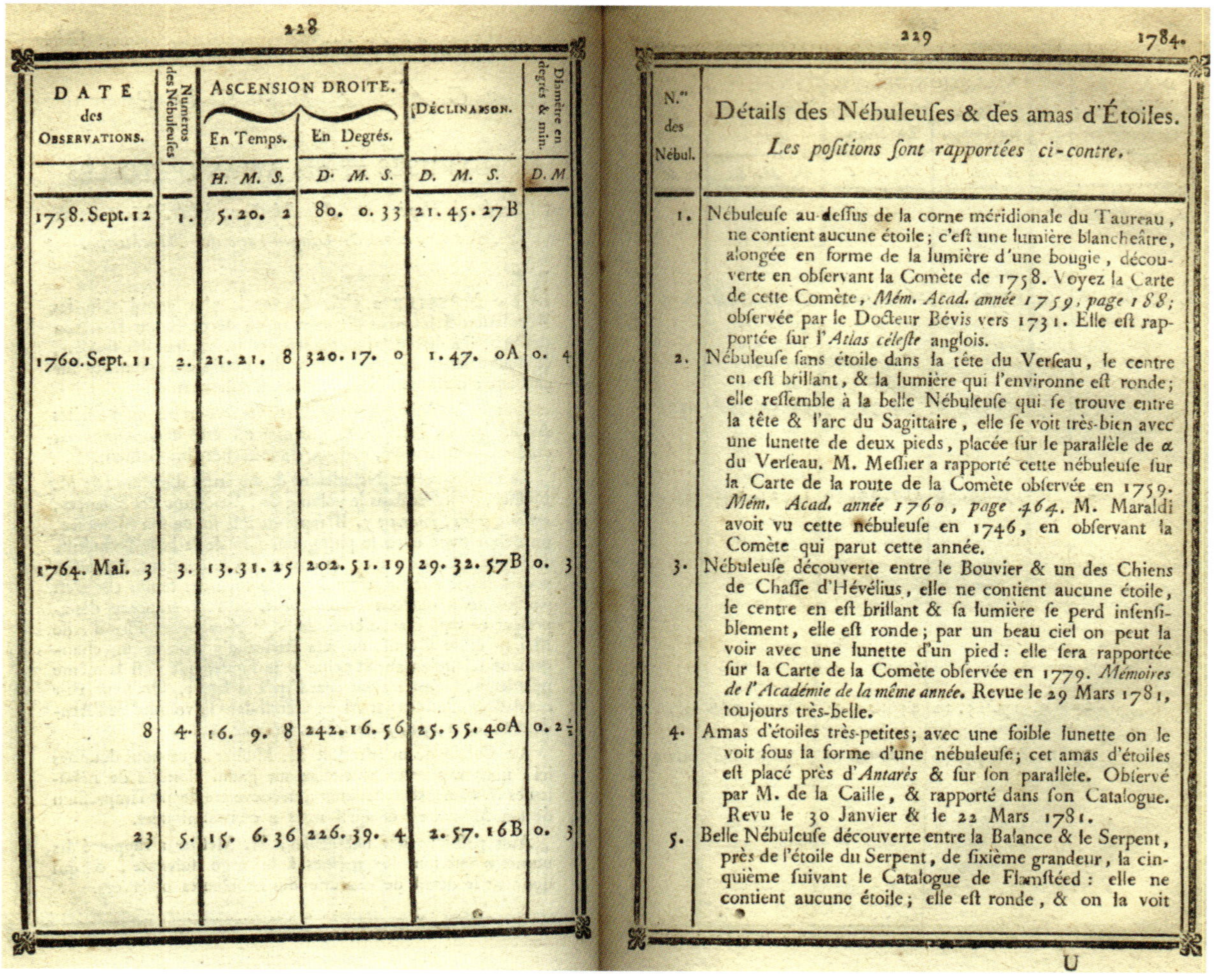

228

DATE des Observations.	Numeros des Nébuleuſes	Ascension droite. En Temps.	Ascension droite. En Degrés.	Déclinaison.	Diamètre en degrés & min.
		H. M. S.	*D. M. S.*	*D. M. S.*	*D. M*
1758. Sept. 12	1.	5. 20. 2	80. 0. 33	21. 45. 27 B	
1760. Sept. 11	2.	21. 21. 8	320. 17. 0	1. 47. 0 A	0. 4
1764. Mai. 3	3.	13. 31. 25	202. 51. 19	29. 32. 57 B	0. 3
8	4.	16. 9. 8	242. 16. 56	25. 55. 40 A	0. 2½
23	5.	15. 6. 36	226. 39. 4	2. 57. 16 B	0. 3

229 1784.

N.os des Nébul.

Détails des Nébuleuſes & des amas d'Étoiles.

Les poſitions ſont rapportées ci-contre.

1\. Nébuleuſe au-deſſus de la corne méridionale du Taureau, ne contient aucune étoile; c'eſt une lumière blancheâtre, alongée en forme de la lumière d'une bougie, découverte en obſervant la Comète de 1758. Voyez la Carte de cette Comète, *Mém. Acad. année 1759, page 188;* obſervée par le Docteur Bévis vers 1731. Elle eſt rapportée ſur l'*Atlas céleſte* anglois.

2\. Nébuleuſe ſans étoile dans la tête du Verſeau, le centre en eſt brillant, & la lumière qui l'environne eſt ronde; elle reſſemble à la belle Nébuleuſe qui ſe trouve entre la tête & l'arc du Sagittaire, elle ſe voit très-bien avec une lunette de deux pieds, placée ſur le parallèle de α du Verſeau. M. Meſſier a rapporté cette nébuleuſe ſur la Carte de la route de la Comète obſervée en 1759. *Mém. Acad. année 1760, page 464.* M. Maraldi avoit vu cette nébuleuſe en 1746, en obſervant la Comète qui parut cette année.

3\. Nébuleuſe découverte entre le Bouvier & un des Chiens de Chaſſe d'Hévélius, elle ne contient aucune étoile, le centre en eſt brillant & ſa lumière ſe perd inſenſiblement, elle eſt ronde; par un beau ciel on peut la voir avec une lunette d'un pied: elle ſera rapportée ſur la Carte de la Comète obſervée en 1779. *Mémoires de l'Académie de la même année.* Revue le 29 Mars 1781, toujours très-belle.

4\. Amas d'étoiles très-petites; avec une foible lunette on le voit ſous la forme d'une nébuleuſe; cet amas d'étoiles eſt placé près d'*Antarès* & ſur ſon parallèle. Obſervé par M. de la Caille, & rapporté dans ſon Catalogue. Revu le 30 Janvier & le 22 Mars 1781.

5\. Belle Nébuleuſe découverte entre la Balance & le Serpent, près de l'étoile du Serpent, de ſixième grandeur, la cinquième ſuivant le Catalogue de Flamſtéed: elle ne contient aucune étoile; elle eſt ronde, & on la voit

U

Pages from Charles Messier's catalogue from 1781, listing Messier objects 1 to 5 and published in 1784 as *Catalogue des nébuleuses et des amas d'étoiles* in *La Connoissance des temps*. The objects listed are: M1, Crab Nebula; M2, globular cluster in Aquarius; M3, globular cluster in Canes Venatici; M4, globular cluster in Scorpius; M5, globular cluster in Serpens Caput.

Charles Messier created his catalogue of nebulae so that the objects would not be mistaken for comets.

Charles Messier

Many catalogues of stars, galaxies and nebulae have been made over the years. The most famous early example was compiled by 18th-century French astronomer Charles Messier (1730–1812). Messier published his *Catalogue of Nebulae and Star Clusters*, containing 110 astronomical objects, in 1781. Of these, 40 are now known to be galaxies. Ironically, Messier was not primarily interested in these objects. He was a comet hunter and compiled his list of fuzzy objects so that he and his fellow comet hunters would not mistake them for comets.

When we refer to these objects in this book, we either use the name Messier in full, or simply the capital letter 'M'. So, for example, the 104th member of this catalogue is referred to as Messier 104 or M104.

William and Caroline Herschel polishing a mirror. Colour lithograph by Alfred Richard Diethe *c.*1896.

William and Caroline Herschel

The famous brother and sister team of William Herschel (1738–1822) and Caroline Herschel (1750–1848) also produced a *Catalogue of Nebulae and Star Clusters*, published in 1786. However, we follow the majority of contemporary astronomers and use Messier's catalogue when referring to individual galaxies. The Herschels lived and worked in Bath in England, and their house has been turned into a museum dedicated to their life and work. They contributed to many other areas of astronomy and have been commemorated in the names of numerous telescopes, prizes and other awards.

Johan Dreyer was born in Copenhagen. In 1874, at the age of 22, he left for Ireland to work on what was at that time the largest reflecting telescope in the world, with its giant 72-inch mirror, the Great Rosse Telescope at Birr Castle, Ireland, known in its day as the 'Leviathan of Parsonstown' (pictured earlier in this chapter). He moved four years later to the Dunsink Observatory in Dublin, and from 1882 he was director of the Armagh Observatory until retiring in 1916.

Johan Dreyer

The most commonly referred to catalogue today is the *New General Catalogue* (NGC), compiled by Johan Ludvig (John Louis) Emil Dreyer (1852–1926) in 1888 while he was director of the Armagh Observatory in Ireland. The library still has a physical copy of Dreyer's original catalogue. We refer to galaxies by their NGC numbers (e.g. NGC 4565) whenever they have one. Where a galaxy has both a Messier number and an NGC number, we use both, with the Messier number first.

Dreyer went on to publish an extension to the NGC in 1895, which he called the *Index Catalogue of Nebulae and Clusters of Stars*, often abbreviated to just the *Index Catalogue* or IC. The second part of the catalogue was published in 1908, adding to the total number of entries by over 5,000.

Portrait of Edwin Hubble, 1931.

Edwin Hubble

A number of galaxy catalogues were compiled in the 20th century, but we use only a few. Most notably, Edwin P. Hubble (1889–1953), whom we will revisit in Chapter 2, spent many years in the first half of the 20th century using the Mount Wilson 60-inch and 100-inch telescopes to photograph, classify and catalogue galaxies.

The fruits of Hubble's labours were published after his death as *The Hubble Atlas of Galaxies* in 1953 by Allan Sandage (1926–2010), based on what Hubble had already completed and his unfinished notes.

This is the body of work that Hubble used to create his galaxy classification system. We use this system throughout this book, with minor additions from later works. However, we note in Chapter 2 that while Hubble's taxonomy is very useful in classifying how a galaxy appears, it should not be confused with any form of evolutionary diagram.

Harlow Shapley with Albert Einstein.

Harlow Shapley and Adelaide Ames

Also in the early 20th century, Harlow Shapley (1885–1972) and Adelaide Ames (1900–32) produced *A Survey of the External Galaxies Brighter than the Thirteenth Magnitude*, which was referred to universally as the *Shapley-Ames Catalog*. It was first published in 1932, and then published in an updated form as the *Revised Shapley-Ames Catalog* (RSA) in 1981 by Allan Sandage and G. A. Tammann. Some updated classifications from the RSA are used in this book. Harlow Shapley was an eminent Harvard professor, and Adelaide Ames was his invaluable research assistant. However, Adelaide Ames died tragically young in a boating accident.

Adelaide Ames, Dr Shapley's assistant at Harvard College Observatory, Cambridge, Massachusetts, from 1924 to 1932, was co-author of the Shapley-Ames catalogue of bright galaxies.

Gerard and Antoinette de Vaucouleurs outdoors at Mount Wilson Observatory, California.

Gerard and Antoinette de Vaucouleurs

In the second half of the 20th century the husband-and-wife team of Gerard de Vaucouleurs (1918–95) and Antoinette de Vaucouleurs (1921–87) worked at the University of Texas, in Austin, and produced the *Third Reference Catalogue of Bright Galaxies*, published in 1991. We referred to this work while writing the book, although we do not refer to the galaxies by their de Vaucouleurs catalogue numbers.

Halton Arp and Barry Madore

The *Atlas of Peculiar Galaxies* was published by Halton C. Arp (1927–2013) in 1966. The galaxies in the catalogue are simply referred to by astronomers as Arp 223, Arp 127 or whatever their catalogue number happens to be. Most of them have other names or previous designations, many with Messier or NGC numbers. The purpose was to bring together in one place a catalogue of galaxies with unusual and peculiar shapes so that astronomers could study these objects as a group and understand them better. Many of the galaxies are the result of collisions, and we discuss some examples of these in Chapter 6.

The *Catalogue of Southern Peculiar Galaxies and Associations* was also published by Arp, this time with Barry F. Madore (b. 1948), in 1987 as a southern hemisphere extension of the original Arp catalogue, with the same aims and objectives. It is usually just referred to as *Arp-Madore*.

Halton 'Chip' Arp (1927–2013) was the author of the *Atlas of Peculiar Galaxies*. In this photograph taken in 1976 he was attending the International Astronomical Union meeting in Cambridge, UK.

Brent Tully and Richard Fisher

In 1987, R. Brent Tully (b. 1943) and J. Richard Fisher (b. 1943) published *The Nearby Galaxies Atlas*, which is universally referred to as the *Tully-Fisher Atlas*. The purpose of their atlas was not to catalogue more objects, but rather to map out all known existing nearby galaxies to aid astronomers in mapping out galaxy clusters and other large structures in the local universe. This was particularly useful when writing Chapter 7.

R. Brent Tully, recipient of the 2014 Gruber Prize in Cosmology, has been an astronomer at the University of Hawaii since 1975. He came to prominence with the publication of a 1977 paper, written with J. Richard Fisher, proposing a relationship between the mass of galaxies and their rotation.

Patrick Moore

Also in the second half of the 20th century, the popular and hugely influential British broadcaster and science writer, Sir Patrick Alfred Caldwell-Moore (1923–2012), presented a series of monthly television programmes for the BBC for over half a century, earning himself a place in *Guinness World Records* for the longest-running TV series in the world with the same presenter.

Sir Patrick Moore in the 1970s, to whom this book is dedicated.

Sir Patrick Alfred Caldwell-Moore, the person responsible for introducing the authors of this book to each other! Seen here with Sir Brian May.

Sir Patrick Moore (right) seen here with author Professor Derek Ward-Thompson (left) and Professor Carlos Frenk on the set of *The Sky at Night*, 2007.

In 1995, Sir Patrick Moore decided to publish a catalogue of objects that were visible with a small telescope or a pair of binoculars, which were often neglected by observers. These objects included nebulae, clusters and galaxies. He could not call it the Moore catalogue, because an 'M' number would be confused with Messier. So he used the first part of his lesser-known hyphenated surname, Caldwell-Moore, thereby giving the objects a 'C' number, in what he called *Beyond Messier: The Caldwell Catalogue*. This contained 109 objects, of which 35 are galaxies.

Michael Koenig and Stefan Binnewies

In 2016, Michael Koenig and Stefan Binnewies published the *Bildatlas der Galaxien* (*Picture Atlas of Galaxies*) in German. This was translated into English by Phillip Helbig in 2017 as *The Cambridge Photographic Atlas of Galaxies*. This proved to be a useful reference guide to the most up-to-date consensus on a galaxy's classification while we were writing this book. However, we have occasionally begged to differ with these authors over the appearance of a few of the galaxies. Any errors introduced in this way are, of course, ours alone.

Others

Other catalogues have been published, and many surveys have been carried out at wavelengths other than the optical (visible light), such as the infrared and the radio, where galaxies can look very different. For example, at the radio wavelength of 21 centimetres, where atomic hydrogen is being traced, the disc of a galaxy can appear significantly larger than the equivalent disc seen in the optical. This is because the optical light traces the stars in a galaxy, but the radio traces atomic gas (hydrogen). The gas stretches to much larger distances, but at densities so low that the gas cannot form itself into stars in the outer parts of the galaxy. Hence the galaxy looks bigger at 21 centimetres.

The *Caltech Observatory List B* (CTB) is a catalogue of radio sources, some of which are galaxies. Another famous early example of a radio catalogue was the *Third Cambridge Catalogue of Radio Sources* (3C), published in 1959, which also contains objects other than galaxies. However, since the current book is mostly about optical images of galaxies, we stick mostly to optical catalogues.

There are other catalogues that are not only of galaxies themselves, but also of clusters of galaxies – which we discuss in Chapter 7. However, these clusters all contain galaxies that are themselves catalogued elsewhere. We mention these cluster catalogues as and when relevant in the text.

Many other works were studied when compiling this book, but those mentioned above were the most influential, both in the selection of galaxies portrayed and in writing the text. However, a most useful book, which we referred to often, was the textbook *Galactic Astronomy* by James Binney and Michael Merrifield, published as part of the *Princeton Series in Astrophysics*. Although not a catalogue, it has become one of the most influential reference works in this field.

Please note that our book is not intended to be a catalogue of galaxies. We have aimed instead to provide an overview of the subject, in words and pictures and stereoscopic images, to communicate the current understanding of these fascinating objects in all their complexity – these billions of vast glowing islands in the immensity of what seems like infinite space and time, the Universe we live in.

NGC 1365

Two observations of a central portion of the galaxy NGC 1365 are combined in this image. The Hubble Space Telescope optical image forms the bottom-left triangle, and the James Webb Space Telescope infrared image is the top right triangle. The galaxy's core is at the centre and its bar structure extends vertically, forming a backwards 'S' shape. In the Hubble optical image, the bar and the beginning of the galaxy's spiral arms are seen as dark dust lanes and some bright blue star clusters. In the Webb infrared image, the dust lanes that cross the bar are now seen brightly. This is because the dust lanes absorb light in the optical (looking dark) and emit it again (by 'shining') in the infrared.

NGC 4945

The galaxy NGC 4945 (or Caldwell 83) lies in the direction of the constellation Centaurus. It is seen almost edge-on, but is nevertheless known to be a barred spiral galaxy. It is around 11 million light years away. It has an active nucleus, and is classified as a Seyfert galaxy.

Building Blocks of the Universe

2

What is a Galaxy?

Far Apart or Close Together?

Space is confusingly large. Stars are a long way apart. If the Sun were shrunk to the size of an orange and placed in London, then Proxima Centauri, the closest star to the Sun, would be in Moscow and slightly smaller, with essentially nothing in between. This is the very essence of isolation.

The picture for galaxies is extremely different. If our own Milky Way Galaxy were shrunk to the size of a 15-centimetre dinner plate and placed at one end of a large household dining table, then the closest galaxy of a similar size, the Andromeda galaxy, would just be at the other end of the same dining table, and it would be about the size of another dinner plate.

This is one of the confusing facts about the size scales in space: galaxies are close together relative to their sizes, whereas stars are a long way apart relative to their sizes. This is why we rarely see stars colliding with one another, but we see many instances of galaxies colliding. We will see examples of what this looks like in Chapter 6, but first we will look at the issue of classifying galaxies.

Galaxy Classification by Shape

The first person to produce a classification of galaxies was Edwin Hubble with the Hubble diagram of galaxies. Hubble was the first to show that spiral 'nebulae', as they were originally known, lay outside the Milky Way and that they were galaxies in their own right. He spent much of his career studying galaxy shapes and types.

The Hubble diagram consists of a number of components. There are spherical and elliptical galaxies, as well as two types of spiral galaxies, ordinary (normal) spirals and barred spirals. Astronomers now call those galaxies that do not fall into any of these types 'irregular'.

The ordinary spiral galaxies are those in which the spiral pattern extends all the way to the centre of the galaxy. The barred spirals have a linear structure across the centre of the galaxy (known as a 'bar'), and the spiral pattern starts at the two ends of the bar and extends to the outer parts of the galaxy.

The Hubble diagram was originally intended to be used as an evolutionary diagram, with galaxies evolving from type to type. However, as noted in the previous chapter, it describes little more than what a galaxy looks like. It is simply a taxonomy of shapes. It does not tell us where a galaxy came from, how it will evolve into the future, or what is happening inside the galaxy today.

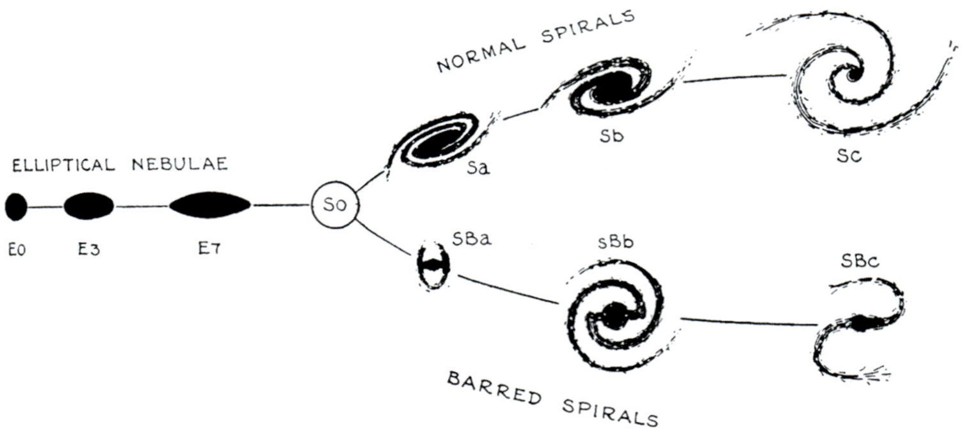

Edwin Hubble's original tuning-fork diagram as published in *The Realm of the Nebulae* (Yale University Press, 1936).

Galaxy classification: the main types of galaxies as originally identified by Edwin Hubble are shown here represented by photographs of actual examples of these types of galaxies. There are also subcategories, as we shall see later.

Edwin Hubble (left) and James Jeans on the Cassegrain platform next to the 100-inch telescope, Mount Wilson Observatory, California. Jeans's theory of nebular evolution preceded Hubble's work.

Galaxy Classification by Colour

The colour-brightness diagram

To understand more about the evolution of galaxies, astronomers have started plotting the colour-brightness diagram, as seen on the next page. This is similar to the Hertzsprung-Russell diagram for stars, which led to so many breakthroughs in the understanding of stars. However, instead of plotting colour against brightness for individual stars, it does this for whole galaxies (see next page).

The vertical axis of each diagram (y-axis) shows the overall colour of the galaxy from blue to red. The horizontal (x-axis) shows the brightness of the galaxy as a whole (also known as its 'luminosity'). So, something dim and blue would lie at the bottom-left of this diagram, and something red and bright would appear top right. The diagram was the first work that showed that spiral and elliptical galaxies have real physical differences beyond their shape.

The interesting thing about this diagram is that almost all galaxies were originally seen to lie in two groups: either in the area marked 'blue cloud' or in the area marked 'red sequence'. Furthermore, the 'blue cloud' group is mostly made up of spiral galaxies, and the 'red sequence' group is mostly populated by elliptical and lenticular galaxies. The gap in between the two groups was labelled the 'green valley', because it appeared that the green area of the plot contained far fewer galaxies. However, our own Milky Way Galaxy lies in the 'green valley', as do some other spiral galaxies.

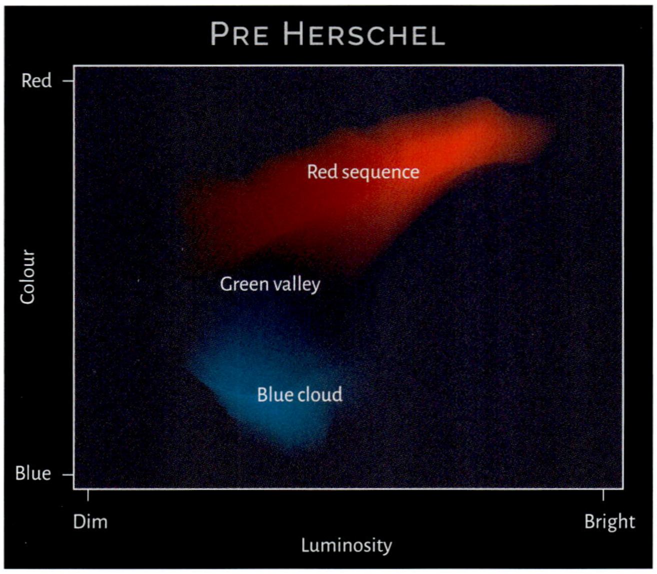

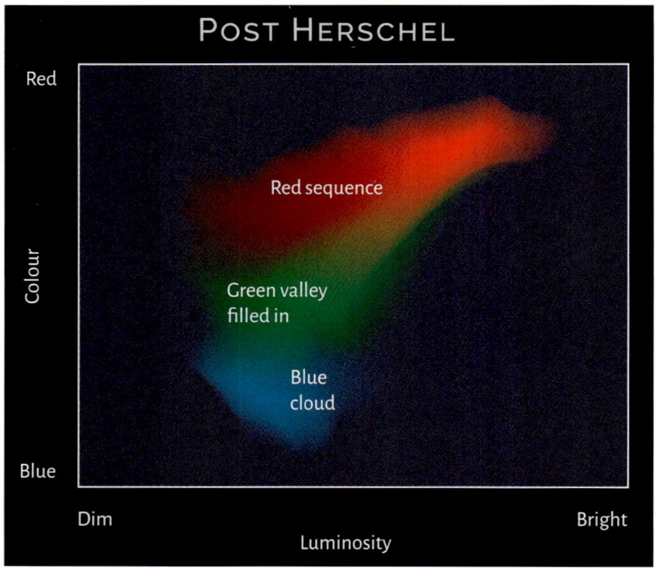

The colour-brightness diagram of galaxies, where brightness runs from left to right and colour runs from bottom to top. The top plot shows the situation before the Herschel Space Telescope, in which astronomers thought that there were two distinct populations of galaxies that sat either in the blue region at the bottom or the red region at the top, with a gap in between, which was called the 'green valley'. The Herschel Space Telescope found that there were galaxies in all areas of this plot and effectively 'filled in' the green valley. This showed that there is simply one continuous population of galaxies of all colours and brightnesses.

How Green Was My Valley?

The Herschel Space Telescope was built by the European Space Agency (ESA) and the long wavelength camera on Herschel was built by a European consortium led by Matt Griffin of Cardiff University in Wales. Furthermore, the team of astronomers who made the observations finding all of the galaxies that populated the green valley was led by Stephen Eales, also at Cardiff University in Wales.

'Blue and live' or 'red and dead'

The colour-brightness diagram can be explained by noting that it is based on images at optical light wavelengths and remembering that the light from a single galaxy is simply the sum of the light from all of its stars. Blue stars are all very young, because blue stars are very short-lived compared to red stars. Blue stars burn through their internal fuel very quickly and die, sometimes in an explosive fashion as supernovae – in some cases after only a few million years. Therefore, all the blue stars that we see must be young.

Red stars can have a variety of ages, depending on their mass, and some of them are very much older than the Sun. Red stars burn their fuel at a much slower rate because, by and large, they are much less massive than blue stars.

For a galaxy to appear blue, it must contain many blue stars. In other words, it must have many very young stars. The only way it can maintain a pool of young stars is by continually forming them. So blue galaxies must be star-forming galaxies.

Conversely, red galaxies must contain almost no blue stars, that is, no young stars. Therefore, such galaxies must have stopped forming stars and become what we refer to as non-star-forming galaxies – or 'red and dead', as astronomers colloquially refer to them. Hence, the colour-brightness diagram tells us what is happening inside the galaxy and whether it is forming new stars or not.

A continuum of galaxies

The Herschel Space Telescope found many more galaxies than were known from optical images taken by the Hubble Space Telescope. These became known as 'Herschel galaxies', and they appeared all over the colour-brightness diagram, as seen on the previous page (lower). This showed that there are not two distinct groups, but rather that galaxies more or less cover the whole of the colour-brightness diagram, filling in the 'green valley'. This was discovered because the Herschel Space Telescope worked in the far infrared, and it happens that the missing 'green' galaxies are bright in the far infrared. Astronomers are still trying to work out why this is the case, but that is why we now know that there is no 'green valley'. Hence, galaxies cannot be simply divided, but rather there is a continuous range of galaxies from blue to red and from faint to bright. We still do not understand how galaxies evolve in the colour-brightness diagram. We will return to Herschel galaxies shortly.

Visible light was all that was available to astronomers for thousands of years, first using the naked eye, and later with optical telescopes. However, astronomers, like all scientists, came to know that visible light is only a very small part of what is known as the 'electromagnetic spectrum', which is the full range of waves that travel through space.

Ground-based radio telescopes began being built in numbers in the 1950s, and then came near-infrared instruments. However, it was necessary to wait for the space age and the advent of satellites working at wavelengths obscured by the blanket of Earth's atmosphere for the rest of the spectrum to become available.

The Herschel Space Telescope, launched in 2009, was the largest infrared telescope ever built. It looked at the sky at far-infrared and longer submillimetre wavelengths. It ceased observing in 2013 after running out of liquid coolant.

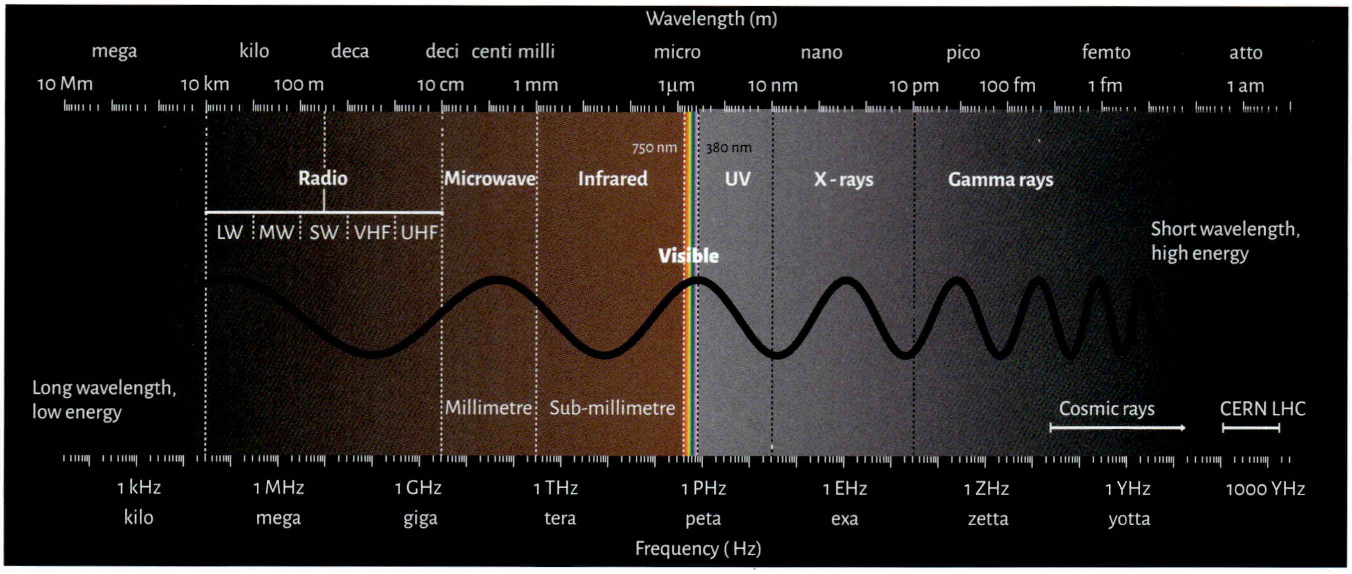

The electromagnetic (EM) spectrum. Both the frequency and wavelength scales are logarithmic – moving from one long tick to the next represents a tenfold increase. As frequency increases, wavelength decreases.

Galaxy Classification by Distance

The most distant galaxies – Hubble's law

Edwin Hubble not only classified galaxies, but also came up with his famous and highly important distance-redshift relation, which we now refer to as Hubble's law. The graph below shows the original data plotted by Hubble, with distance measured on the horizontal (x-axis) and redshift vertically (y-axis). Redshift is a measure of how far the light from a galaxy has been moved towards the red end of the spectrum. The more a galaxy's spectrum is redshifted, the faster it is receding (see box on next page).

Hubble drew a straight line through the points on his graph, showing that distance was related to redshift. Consequently, he formulated his law stating that galaxies further away from us are receding faster than those closer to us in proportion to their distance. The average ratio (constant of proportionality) is known as Hubble's constant. The more the light from a galaxy is redshifted, the longer the wavelength of electromagnetic radiation at which it emits most strongly. This led astronomers to realize that the most redshifted galaxies would be brightest in the infrared rather than the optical, with extremely redshifted galaxies being seen in the far infrared and even microwave regimes. Astronomers began to use the Herschel Space Telescope in the far infrared at wavelengths roughly from 0.1 to 0.5 millimetre, and the longer wavelength James Clerk Maxwell Telescope (JCMT), which operates roughly from 0.75 to 2 millimetres, to study the most redshifted, and hence the most distant, galaxies.

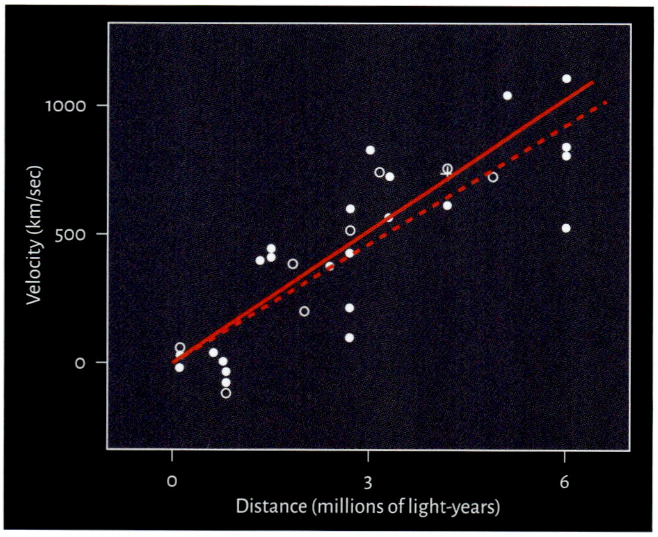

Hubble's data, which he plotted in 1929, led him to deduce a relation between the distance to a galaxy and the speed at which it was moving away. Note that the velocities were measured by Vesto Slipher (1875–1969) at the Lowell Observatory in Flagstaff, Arizona.

Redshift

Redshift (or Doppler shift – see Chapter 5) is an effect where something moving away from you has its light moved to longer (redder) wavelengths. The faster an object is moving, the greater the shift in wavelength. This happens because as the object moves away, the waves of light it emits get stretched. This is similar to the way that the tone of a siren changes as it moves away from you. Edwin Hubble found that all distant galaxies are moving away from us, and in fact the furthest-away galaxies are moving fastest. So the furthest galaxies have their light shifted to the longest wavelengths. Note that the galaxies are moving away from us because all of space is expanding.

Electromagnetic (EM) Spectrum

Optical (or visible) light occupies a very small part of what we call the electromagnetic spectrum. This is the full range of waves from short wavelength, high-energy X-rays, to long wavelength, low-energy radio waves.

Galaxy Classification by Wavelength

Herschel galaxies

In the early 2000s, the Herschel Space Telescope flew. Naturally astronomers looked for redshifted galaxies at the wavelengths of Herschel, typically at 0.1 to 0.5 millimetres.

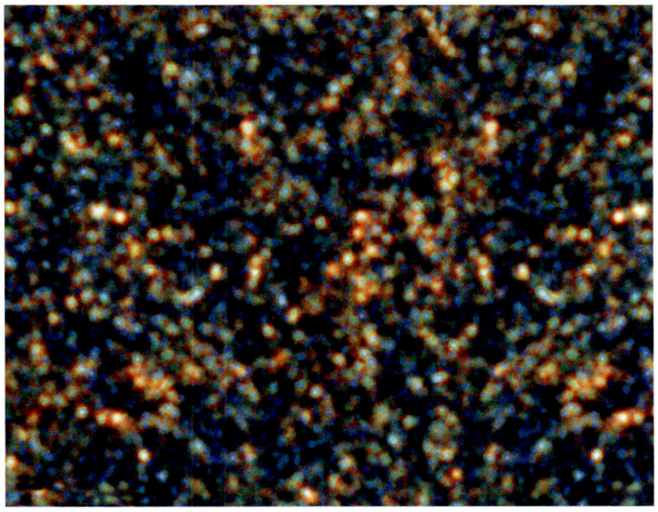

A far-infrared picture of the most distant galaxies in the Universe, taken by the Herschel Space Telescope at wavelengths of 0.25, 0.35 and 0.5 mm, shown in false colours as red, green and blue, then put together in a composite image.

The image on this page was taken in the far infrared by the Herschel Space Telescope. It is made from three different far-infrared wavelength images, put together in a red-green-blue false-colour image. It is full of different coloured spots, and each one of these is a very distant galaxy. They are so far away that we cannot resolve any structure in them, but they are believed to be the very early precursors, or counterparts, of the galaxies we see in the nearby Universe today.

Each one of these galaxies is seen at an apparent distance of 10–12 billion light-years, when the Universe was about one-tenth of the age it is today. They are called 'Herschel galaxies' after the telescope that found them. We know that the Herschel galaxies are a long way away because they are so bright in the far infrared. Because of the redshift effect (described on the previous page), Herschel sees optical light from galaxies a very long way away that has been redshifted to far-infrared wavelengths.

False-Colour Images

Our eyes are only able to see a small amount of the electromagnetic spectrum – the part that we call 'visible' or 'optical' light. We cannot see the rest of the spectrum, such as X-rays, gamma rays, ultraviolet, infrared or radio waves with our eyes. To show wavelengths in invisible parts of the spectrum, astronomers use what are called a 'false-colour images'. These show the longest wavelength infrared image in red, the middle wavelength image in green and the shortest wavelength image in blue. This conveys wavelength to us in a manner our eyes can see and our brains can interpret.

Resolution

Resolution is a property astronomers use to describe how small a detail on an object can be seen by a given telescope. The bigger the telescope, the better the resolution at a given wavelength that is being detected. The longer the wavelength, the worse the resolution for a particular size of telescope (strictly, it is the diameter of the telescope that matters). For example, the Herschel Space Telescope is smaller than the James Webb Space Telescope, and it works at longer wavelengths, so Herschel has poorer resolution. Where JWST sees a spiral galaxy, Herschel sees only a point of light

SCUBA galaxies

The James Clerk Maxwell Telescope (JCMT) in Hawaii has had a sequence of cameras that operate at 0.8 millimetres. The first such camera was known by the rather unwieldy title of the Submillimetre Common-User Bolometer Array, or SCUBA for short (it was based in Hawaii, after all!). Note that a bolometer is just a detector of infrared radiation, similar to how an individual pixel in the camera in your phone is a detector of visible radiation.

The JCMT discovered, in the 1990s, the first few of the most distant galaxies, whose light had been redshifted all the way to microwave wavelengths. They were observed at a wavelength of 0.8 millimetres and were known as 'SCUBA galaxies', after the camera that detected them. It was soon realized that the JCMT was detecting only the most massive galaxies, known as giant elliptical galaxies, but that otherwise they were much like galaxies we see in the more nearby Universe.

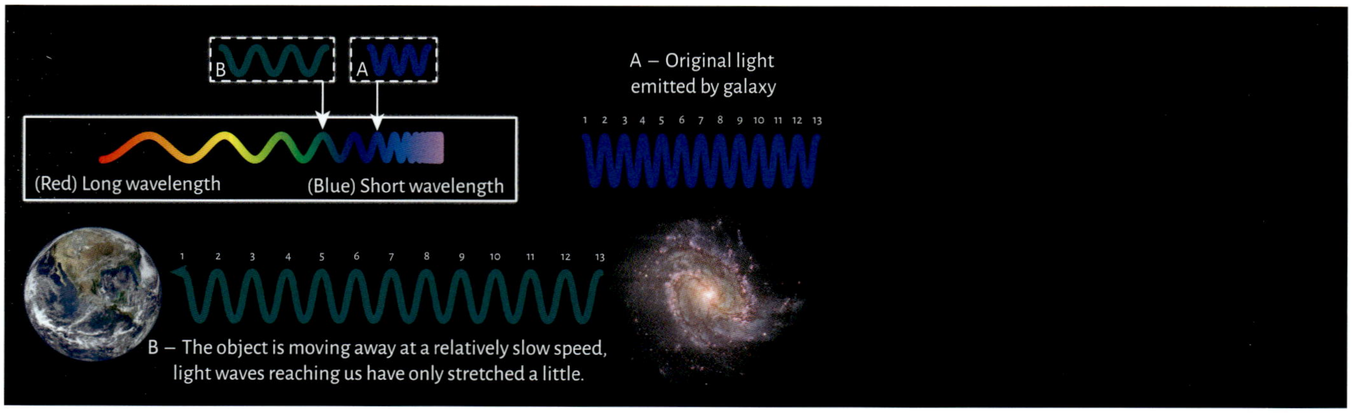

Light from a relatively nearby, slow-moving galaxy is redshifted (stretched) least.

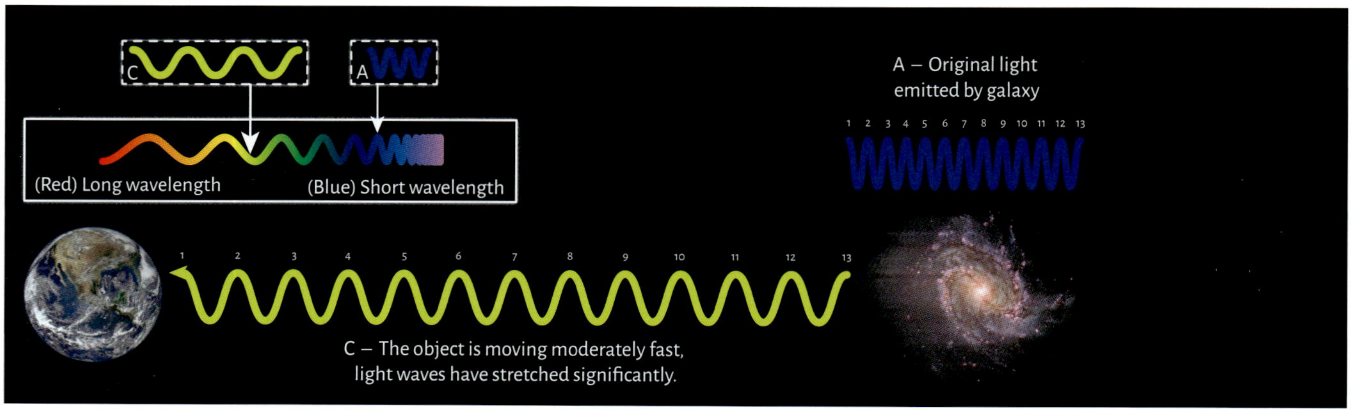

Light from a more distant, faster-moving galaxy, is redshifted more.

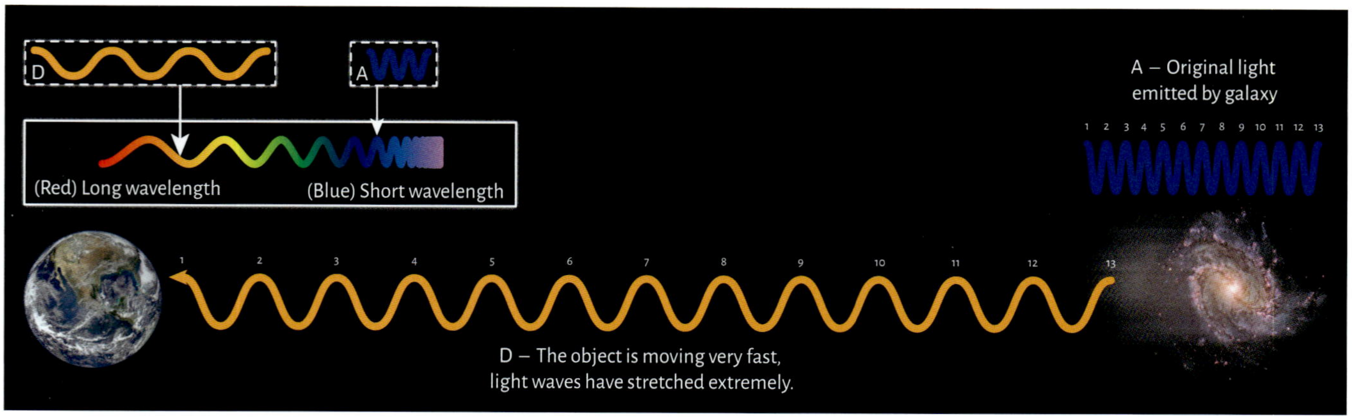

Light from a remote, fast-moving galaxy has the greatest redshift.

James Clerk Maxwell (1831–79)

James Clerk Maxwell was a remarkable 19th-century physicist. He was born in Edinburgh into a relatively wealthy family and grew up on the family country estate in the west of Scotland. He studied at Edinburgh and Cambridge universities and taught in Aberdeen and at King's College, London. He made many contributions to science, but the greatest of these was being the first person to discern the true nature of light. He was a brilliant mathematician who put together a set of previously known equations and showed that together they proved that light is an electromagnetic wave. He thereby discovered what had puzzled humankind since antiquity, namely the true nature of light. Among physicists, Maxwell is regarded as being as important as Newton.

James Clerk Maxwell, aged 24 in 1855, presenting his work 'Experiments on Colour' to the Royal Society of Edinburgh. He is holding one of the colour wheels that he used to mix colours.

The James Clerk Maxwell Telescope sits at 4,092 metres on top of Mauna Kea, an extinct volcano on the Big Island of Hawaii in the Pacific Ocean.

Webb galaxies

In 2022, the James Webb Space Telescope (JWST) captured a very deep near- to mid-infrared image, known as the JWST Deep Field (JWDF). In this image we are seeing some galaxies when the Universe was only about 5 per cent of its current age. We are looking back 95 per cent of the way to the beginning of the Universe.

The James Webb Space Telescope (JWST): NASA technicians moving the telescope inside a clean room at the NASA Goddard Space Flight Center in Greenbelt, Maryland, before launch. The telescope's 18-segmented gold mirror has captured the most detailed images ever in infrared light from the first galaxies that formed in the early Universe.

A European Space Agency (ESA) Ariane 5 rocket launched the James Webb Space Telescope on 25 December 2021.

The James Webb Space Telescope Deep Field (JWDF) shows an image of some of the most distant galaxies ever seen, from a time when the Universe was only a fraction of its current age. The appearance of some galaxies as very long and thin is the result of what is known as gravitational lensing, where light from distant galaxies appears to be distorted by the gravitational effects of nearer galaxies that act to curve space.

As can now be seen at the higher resolution of JWST, the galaxies in the JWDF appear no different from the elliptical and spiral galaxies that we see in the less distant Universe around us today. So astronomers are left with a conundrum: if galaxies have always looked the same, do they evolve at all, and if they evolve, how does this evolution manifest itself?

Galaxies in Context

The inflationary Big Bang model

The Universe is believed to have begun in a 'fireball' known as the Big Bang. Immediately after that, the Universe expanded very quickly during a period known as 'inflation'. During this time, areas of the Universe that were close together and similar to each other were spread over vast distances across the Universe, which explains why regions of the Universe that are now a long way apart look similar to each other.

The History of the Universe

A diagrammatic representation from the Big Bang to the present day.

10^{-36}–10^{-32} second

Inflation: after the Big Bang, it is postulated that the Universe expanded at an unimaginably rapid rate; this is necessary to account for the appearance of the Universe as we see it today.

10^{-6} second

Quarks combine to form protons and neutrons, the building blocks of atoms.

1 second

As the Universe cools, fundamental particles called leptons form; these include electrons, which are essential to the formation of atoms. Electrons collide with light particles, photons, and no light can propagate. The Universe remains opaque.

3 minutes

It is now cool enough for protons and neutrons to stick together to form the nuclei of hydrogen and helium atoms.

Zero to 10^{-43} second

The Planck time. Nothing is known about this era.

380,000 years

:ool enough to allow hydrogen and
um nuclei to capture electrons, and
m neutral atoms. Light can finally
pagate and the Universe becomes
nsparent. The cosmic microwave
ckground (CMB) radiation that we
etect today is that very 'first light'.

200 million years

Gas condenses under gravity and the first stars ignite, and the galaxies appear from the 'lumps and bumps' in the CMB.

400 million years

Around this time the stars clump together to form galaxies.

13.8 billion years

The present day. With modern telescopes we are able to look back to the time the CMB was formed.

This accords with the predictions of the 'inflationary Big Bang model', which is our current model to explain the evolution of the Universe. This similarity everywhere leads to one of the underlying principles of cosmology, known as 'homogeneity', which is simply that the Universe appears the same wherever you look – if you look on large enough scales. This was illustrated rather beautifully by the James Webb Space Telescope when it captured the Deep Field image.

The cosmological principle

It seems astounding that the Universe contains clusters of galaxies and super-clusters of clusters and even webs of superclusters that appear homogeneous all over the Universe. But enough pictures have been taken by many telescopes in every direction in the sky for us to believe that the Universe looks very similar in all directions on a large enough scale. This is known as being 'isotropic', meaning the same regardless of direction or orientation. The scale that is defined as 'large enough' is currently a matter of debate among astronomers, but it must be at least billions of light-years.

The two facets of the Universe, being isotropic and homogeneous, together are known as the 'cosmological principle'. Along with general relativity, this is the underlying principle for the study of cosmology on which all of our understanding of the Universe rests. This is currently under question – see Chapter 7.

The cosmic microwave background

When the Universe was very young, it was made of nothing but free-flying particles, such as protons and electrons. Light cannot travel through such a medium, as the individual photons that make up light interact with the particles and are scattered away out of our line of sight. Trying to look back before this time is rather like looking through frosted glass.

After about 380,000 years (a very short time in astronomy), the Universe cooled down enough for the protons and electrons to join together to form atoms. Only the two simplest atoms, hydrogen and helium, were formed at this time. Following this, light could travel unimpeded through the Universe. This point in time is known as the last scattering of light, and it is best seen at microwave radio wavelengths. The radiation from this time is referred to as the cosmic microwave background (CMB). The crucial thing about the CMB is that it is a snapshot in time that gives us a key to what went before.

The reason that the CMB is so crucial is that up to that point, radiation and matter were inextricably linked together because of the free-flying particles interacting with the light. Therefore, the distribution of matter was indelibly imprinted into the CMB, which preserves that imprint to this day.

Bell Labs' Horn Antenna in New Jersey. In 1964, while using the Horn Antenna, Arno Penzias (right) and Robert Wilson (left) stumbled on the cosmic microwave background radiation that permeates the Universe.

The Cosmic Microwave Background (CMB)

The cosmic microwave background represents the furthest point in the Universe that can be seen by any telescope at any wavelength, however powerful it may be. Beyond this point, light cannot travel in a straight line through the Universe. Before this time the Universe is filled with free-flying protons and electrons, which scatter light and regularly randomize its direction. The effect of these constant scatterings was to make the Universe opaque. After this point the protons and electrons join together to form atoms, and light can then travel freely through a sea of atoms. The CMB was originally extremely hot, and emitted at short wavelengths, but the effect of 'redshift' has taken it from short wavelengths to long wavelengths and we see it today as microwaves.

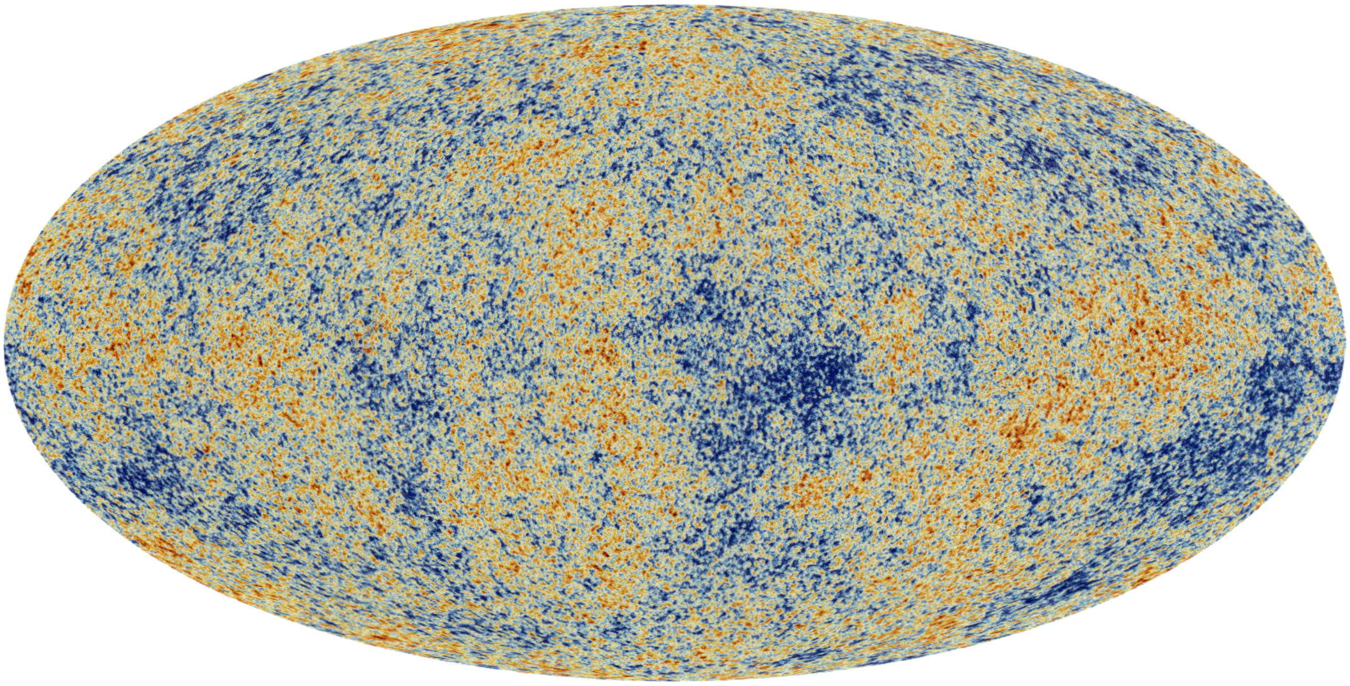

The cosmic microwave background. Red and yellow indicate regions that are slightly warmer than average, while blue regions are cooler than the average.

The image of the CMB (previous page) shows a structure that is not at all uniform on smaller scales. There are brighter spots seen in red and yellow, and there are fainter spots seen in light and dark blue. The bright spots coincide with where the matter was most dense, and the faint spots with where the matter was least dense. These bright and faint spots are known as 'inhomogeneities' in the CMB. They represent the 'lumps and bumps' in the matter.

After about another 200 million years the first stars and galaxies appear out of these lumps and bumps, as seen in the JWDF (earlier in this chapter). We now look at how this came about.

Dark matter

We now know that there is far more matter in the Universe than even the Hubble Space Telescope or the James Webb Space Telescope can see. The matter that we can see, which we might call normal matter, is what scientists call 'baryonic matter'. The extra matter is known as 'dark matter', and it has been shown that it must have played a pivotal role in forming galaxies in the early Universe.

This then raises the question, 'what is dark matter?' The simple answer is that astronomers do not know. It may consist of more exotic and unusual particles than we know about. The problem is that it does not interact with electromagnetic radiation in the way that baryons such as protons and neutrons do, so it is impossible to see it or test it in the laboratory.

However, there is one way in which dark matter does interact with normal matter, and that is through gravity. It seems that dark matter has mass. This is how we know it is there. We can observe the effect of the dark matter from its interaction with the matter that we can see.

A big surprise came when astronomers were able to measure the rotation of the Milky Way Galaxy as a whole. They discovered that the stars in the outer parts of the Galaxy moved at roughly the same speed as those in the inner part. This is known as a 'flat rotation curve', that is, the speed remains the same, or 'flat', across the Galaxy. We now know that all spiral galaxies like the Milky Way have the same flat speed distributions.

The reason astronomers were surprised was that if what we can see is all the matter there is in the Galaxy, then we would see orbital speeds decreasing as we move outwards in the Galaxy – as shown by the dashed curve in the diagram on the next page.. This dashed curve is close to what is known as a Keplerian rotation curve and is what would be predicted if there was only visible matter in the Galaxy.

In the Solar System, roughly 99% of the mass is right in the middle – in the Sun itself. In the Galaxy, however, the visible matter is spread out into a disc, so it should not display exactly Keplerian rotation. Nevertheless, there is far more visible matter near the centre than at the edge, so the curve should be close to Keplerian, and not flat.

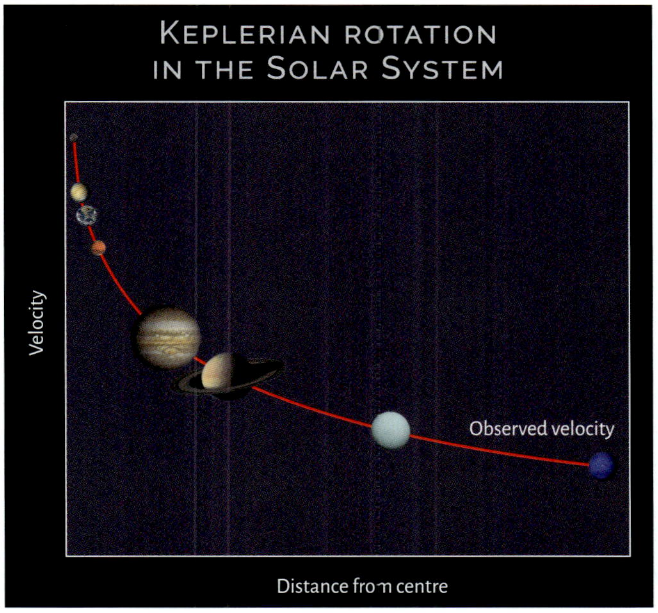

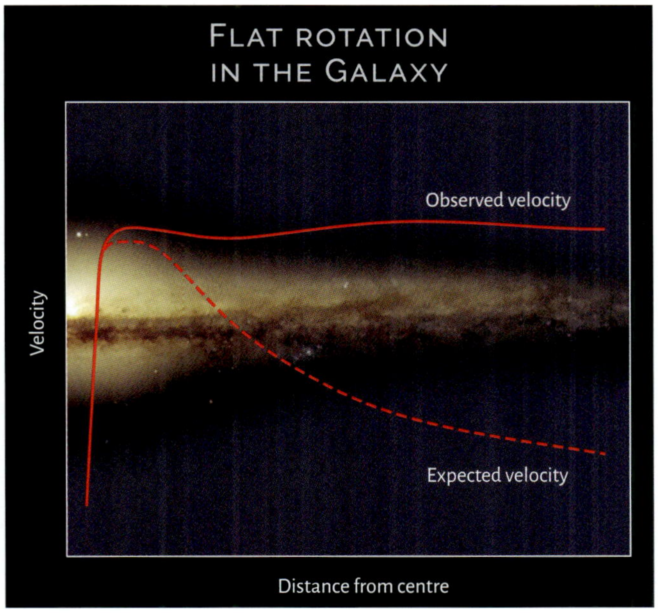

Graphs of the rotation velocity around the centre of the Solar System (left) and the Galaxy (right), plotted against distance from the centre along the bottom. The Solar System has a Keplerian rotation curve, whereas the Galaxy shows a flat rotation curve (solid line). The dashed curve for the Galaxy is what we would see if there was no dark matter, and it is similar to the Keplerian curve.

Keplerian Rotation

Within our Solar System, the planets nearer the Sun, such as Mercury, Venus and Earth, travel in their orbits much faster than those further from the Sun, such as Jupiter, Saturn and Uranus. This fact was noted by Kepler, and we now refer to the collective motions of the planets as obeying Keplerian rotation.

Baryonic Matter

The 'normal matter' that we know about consists of protons and neutrons, among other so-called elementary particles that make up the atoms and molecules that constitute our everyday world. These protons and neutrons are collectively known as 'baryons', and 'normal matter' is sometimes known as 'baryonic matter'.

This montage depicts the collision between two massive clusters of galaxies known collectively as MACS J0018.5. As the collision progresses, the dark matter in the galaxy clusters (its presence indicated by blue) moves ahead of the associated clouds of hot gas, or normal matter (orange). This happens because, while both dark matter and normal matter feel the pull of gravity, only the normal matter experiences additional effects like friction, shocks and turbulence, which slow it down during the collision. The blue areas have been identified as regions where gravitational lensing is occurring (see caption to the James Webb Deep Field image earlier in this chapter) and hence is where the dark matter is concentrated.

The fact that the galactic curve is flat means that the total mass of a spiral galaxy cannot be concentrated near the centre. To produce such flat distributions of speed, the mass must be spread out into a giant sphere or 'halo' around the disc of the spiral galaxy. In addition, there must be far more dark matter than the normal baryonic matter that makes up all the stars and planets within a galaxy that we can see. This came as quite a surprise when it was discovered.

The Formation of Galaxies

The inhomogeneities ('lumps and bumps') in the matter of the Universe, which are reflected in the CMB, expanded with the general continuing expansion of the Universe and became enlarged. The largest inhomogeneities became the sizes of what we now see today as clusters of galaxies - they were truly massive.

The majority of the mass was in the form of dark matter, as discussed above. In each case, once inflation had stopped, the mass was unstable and began to collapse. As it collapsed, it fragmented into objects the size of galaxies. Each of these galaxies began to collapse in onto itself to create the galaxy shapes we see today.

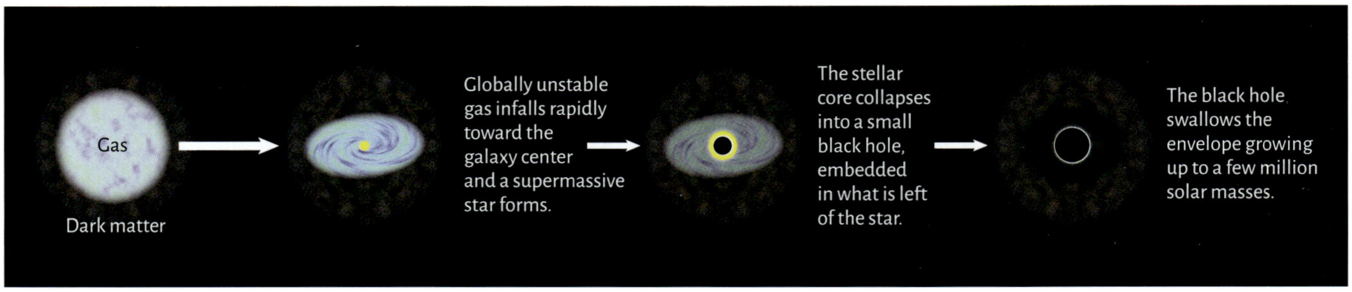

One theory of how a galaxy forms with a black hole in its centre.

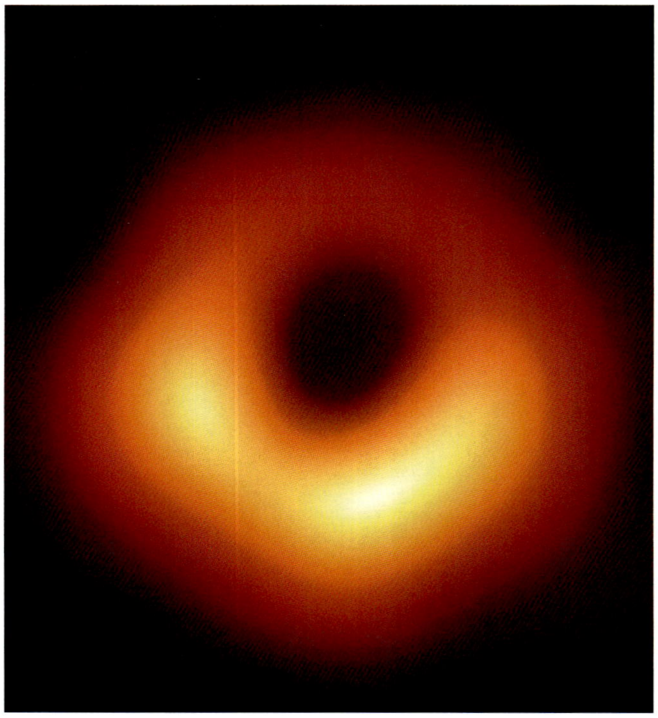

The first black hole ever to be imaged. This is the black hole at the centre of the galaxy Messier 87, and it is catalogued as M87*.

Supermassive black holes

The centre of each galaxy is where the most mass collects. It is said to be at the bottom of a massive gravitational potential well. Here an enormous star starts to form, which can be a million times more massive than the Sun. This star does not have time to evolve in the way of a normal star.

In quite literally a flash, it implodes into what we call a black hole, from which nothing, not even light, can escape. The black hole carries on growing and can become a few billion times more massive than the Sun. Meanwhile, the rest of the galaxy continues to orbit the black hole at a much greater distance than the event horizon (see Chapter 3).

This is what happened in the centre of the galaxy Messier 87. In 2019 astronomers released a picture of the so-called 'event horizon' around this black hole. The image generated a great deal of interest from both astronomers and non-astronomers.

The black hole itself is seen as a dark patch at the centre of the image. The ring shape that is seen is a combination of matter orbiting and matter seen behind the black hole. The bottom of the ring is foreground. This image was taken by the Event Horizon Telescope (EHT) Collaboration. One of us (Derek Ward-Thompson) is a member of this collaboration.

Active galaxies

Most, if not all, galaxies are believed to have these supermassive black holes in their centres. The evidence for many galaxies having black holes lies in what astronomers term 'active galaxies', or 'active galactic nuclei' (AGN). Active galaxies manifest themselves in many ways, but one thing that all have in common is that they eject matter in some form or another from their centres, or nuclei. These ejections of matter are referred to as 'jets' or 'outflows'.

Jets

The galaxy M87 has a powerful jet emerging from its centre. Such jets are composed of extremely energetic matter at high temperatures that is highly ionized (electromagnetically charged) and travelling at a significant fraction of the speed of light – often more than 90%. The M87 jet is so fast and long that it has entirely escaped the galaxy in which it originated. To give the scale of such objects, the picture (previous page) showing the M87 central black hole and its surroundings is only a few light-days across; the images of M87 showing the jet are a few hundred thousand light-years across. Yet it is believed that the black hole is responsible for launching the jet.

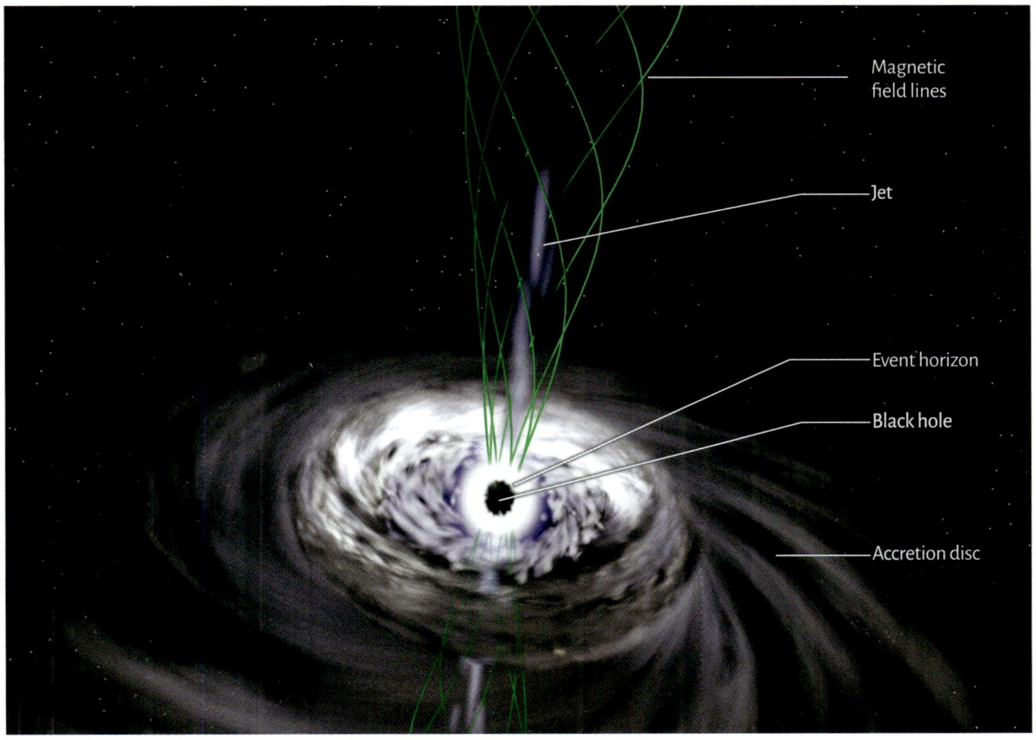

A schematic of a black hole at the centre of a galaxy launching a jet.

What is generally believed to be happening is that matter is spiralling in towards the black hole in a disc, which is known as an 'accretion disc' – looking a little like the rings of Saturn, but on a much larger scale. There is also a very strong magnetic field being twisted around in the vicinity of the black hole, in a helical, or corkscrew-like structure of its own.

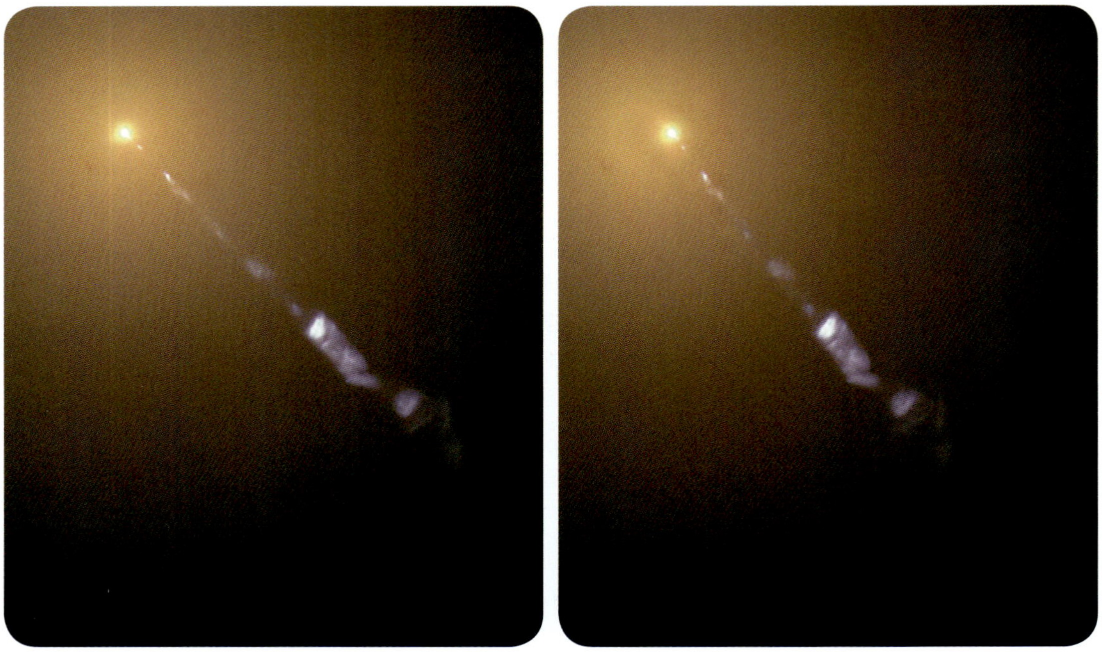

Hubble Space Telescope image of the jet from the black hole in M87. Note how the jet extends across space for a distance many times the size of the galaxy.

Magnetic fields and charged particles interact very strongly, so the matter that is charged, which is spiralling in, is greatly influenced by the magnetic field, and some fraction of the matter does not make it into the black hole. Instead, it is flung out by the corkscrew-shaped magnetic field along the axis of the disc. This matter is held close to the axis, confined by the magnetic field. The matter has so much energy as it is falling into the disc that it is ejected at extremely high velocity, roughly perpendicular to the plane of the galaxy, so it is very soon outside the galaxy in what we call 'intergalactic space'. This space is extremely rarefied – even more so than the space between the stars within a galaxy. Consequently, the jet keeps going for huge distances – there is quite literally nothing to stop it.

Is the Milky Way an active galaxy?

It is believed that most disc galaxies have black holes at their centres, but not all of them are active galaxies. Our own Milky Way Galaxy has a black hole at its centre, but it is not classed as an active galaxy. Astronomers have explained this by realizing that galaxies have phases when they are active and phases when they are quiet, or 'quiescent'.

The question about our Galaxy was answered when gamma-ray and X-ray satellite telescopes were pointed towards its centre. What was seen in gamma rays was two large round areas of emission. These are believed to be bubbles that have been blown in space, one on each side of the centre of the Galaxy, above and below the plane of the Milky Way. Furthermore, X-ray emission was detected along the edges of the bubbles.

The bubbles appear to have been blown out by a previous episode in which our Galaxy was active and had matter flowing out from its centre, 'blowing the bubbles'. This outflowing matter now appears to have stopped, leaving the bubbles as evidence of previous activity.

Activity switched off

The reason why the activity has switched off became apparent when astronomers studied the centre of our Galaxy. They found that there is a small amount of matter in the immediate vicinity of the black hole itself, and around this is a cavity a few light-years across that has a very low density. This is called the 'inner cavity'.

Therefore, the lack of any significant amount of matter in the cavity means that there is very little or no matter falling into the black hole, and hence there is no matter being ejected in the form of jets or outflows.

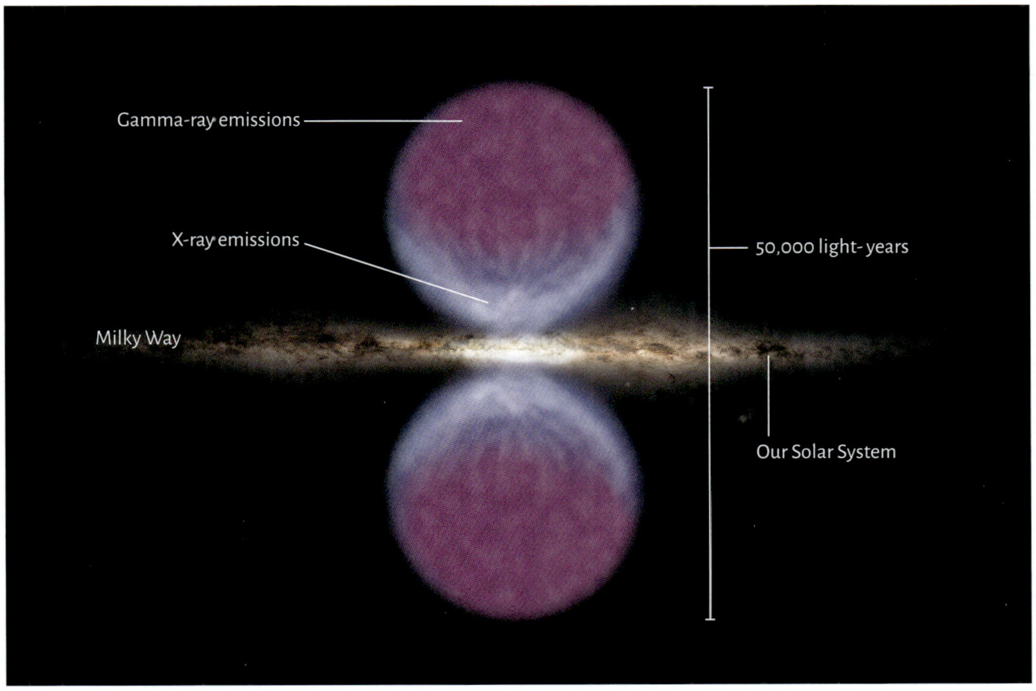

A schematic of our own Milky Way Galaxy at a compilation of wavelengths. An optical image of the Milky Way is seen across the centre of the image. Overlaid on this in purple is an image at gamma-ray wavelengths showing two 'bubbles', one each side of the galactic plane. Overlaid in blue is an X-ray image outlining the edges of the bubbles.

The Inner Cavity

The inner cavity of the Milky Way may not contain much gas, but it does include a small number of stars whose positions have been plotted with great accuracy over a number of decades. They have been shown to be orbiting a point that emits no detectable radiation, that is, the black hole at the centre of the Galaxy.

If all galaxies pass through phases of having a large amount of matter in their centres and then phases with cavities around their black holes, this would explain why only a relatively small fraction of known galaxies is active – even though most are believed to have black holes in their centres.

The Milky Way

This beautiful view of the heart of the Milky Way is a stunning backdrop to the Chajnantor plateau in the Atacama Desert of Chile, one of the highest and driest habitats on Earth, and home to the Atacama Large Millimeter/submillimeter Array (ALMA) telescope that is studying light from some of the coldest objects in the Universe. This photograph was captured with an optical camera.

Our Home Galaxy

3

The Milky Way

An Insider's View

We live in a galaxy that we call the Milky Way, and sometimes we just have to stand in awe of its beauty. Our viewpoint of the Milky Way is from the inside, so we cannot see what our Galaxy looks like from the outside. However, using the computing power that is now available to us, together with novel ways of handling images, we can obtain a reasonable idea of its structure.

The Milky Way is a spiral galaxy, and our Solar System sits in roughly the midplane of the disc, about two-thirds of the way from the Galactic Centre to the edge. As we look up from our viewpoint on Earth, we see the disc of the Galaxy edge on, from within the disc – providing we have a dark enough sky.

The pictures of our Galaxy in these pages show its true beauty and splendour. The dark lanes that we can see in the disc roughly delineate the midplane of the Galaxy. If we lock in the right direction, towards the constellation of Sagittarius, we can see the bulge in its centre.

We can use these images to start building up a picture of what our Galaxy might look like from the outside. In Chapter 4 we give some examples of external galaxies that appear to resemble what we see.

However, our unique viewpoint within our own Galaxy gives us an insight into its structure that we cannot match for any other galaxy, purely because of the distances involved. Our close-up look at our Galaxy allows us to infer what is happening in other similar galaxies elsewhere. We can look for pattern similarities in structures and work out what is happening in other galaxies as a result of our discoveries in our own.

Consequently, the study of our Galaxy, while being interesting in and of itself, is important for our studies of all galaxies. Ours may be only one of hundreds of billions of galaxies, but studying it gives us more detailed insights than we could obtain by any other means into the astronomy and astrophysics of all galaxies, in both their similarities and differences.

An amazing stereoscopic conversion by J.-P. Metsävainio of the Milky Way Galaxy as seen from Earth enables us to clearly understand that the dark areas are not due to a lack of stars but to obscuration by dark nebulosity.

The image on the next page shows a view from outside the Galaxy constructed by carefully measuring distances to all components of the Galaxy. The Milky Way looks like a classic barred spiral galaxy, with a number of what are known as 'spurs' apparently radiating away from the main spiral arms. Each of the features in this map is named according to the direction in which it lies if projected onto the night sky, and whichever constellation happens to lie in that direction (for instance, the Perseus Arm). The Sun is marked and can be seen between spiral arms.

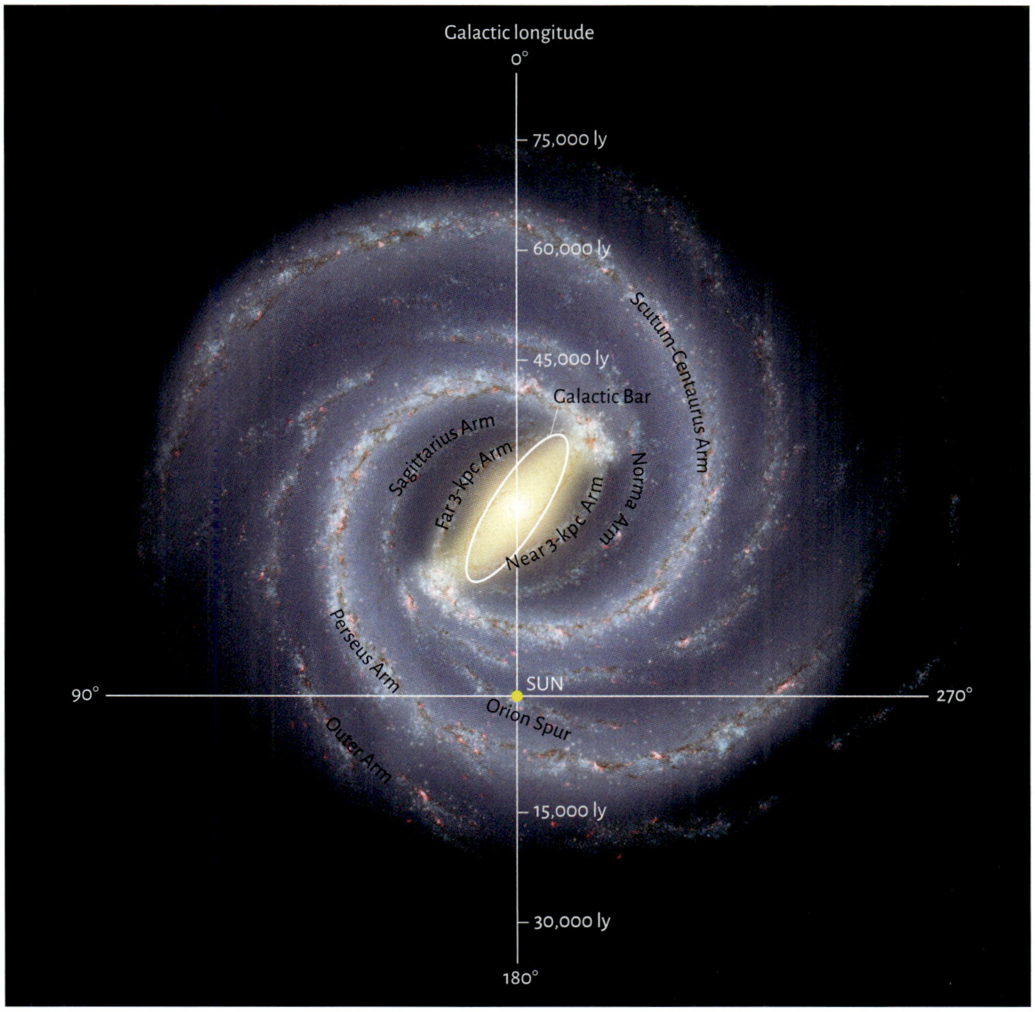

An artist's impression of the Milky Way seen from above. The Galaxy's major arms (Scutum-Centaurus and Perseus) can be seen attached to the ends of a thick central bar, while the two minor arms (Norma and Sagittarius) are less distinct and located between the major arms. The major arms consist of the highest densities of both young and old stars; the minor arms are mainly filled with gas and pockets of star-forming activity.

The Milky Way's Interstellar Medium (ISM) Gas

The spaces between the stars in the Milky Way are not empty. They contain, mainly at very low densities, grains of dust and gases of various kinds.

Atomic gas

Atomic gas consists mostly of hydrogen and helium atoms, with some heavier elements. Hydrogen atoms emit strongly in the radio part of the spectrum at a wavelength of 21 centimetres. The images on the next page show galaxy M101, or NGC 5457, at optical wavelengths next to the same spiral galaxy at 21 centimetres. Almost all of the spiral arm structure can be seen in both images. This shows that the atomic gas is spread throughout the disc of the galaxy in the same way as all the other constituents of the galaxy. The big difference is seen at the centre, where the optical image is very bright, but the 21-centimetre image is dark. This is because the density towards the centre of the galaxy is much higher, and at high densities the atoms in the gas can join together to form molecules and what is known as 'molecular gas'. This can happen in other parts of a galaxy too, wherever the density is high enough. However, throughout the disc of the galaxy, the atomic hydrogen traces the spiral structure and is therefore well intermingled with the other constituents, such as the stars and the rest of the ISM.

The left-hand picture shows a black-and-white optical photo of the spiral galaxy M101, or NGC 5457. The right-hand image shows the same galaxy taken in the 21-cm radio wavelength of atomic hydrogen, a part of the ISM, showing that the atomic gas follows the same spiral structure as the other components of the galaxy.

Molecular gas

Molecular gas includes simple molecules, such as molecular hydrogen (H_2), carbon monoxide (CO), silicon dioxide (SiO_2) and many others, but also contains some very complex molecules such as carbon-chain molecules. The JWST picture of part of Ophiuchus (next page) shows molecular gas in a cloud. The molecular gas shows up as dark red and black in the image. It is dense, so molecules have formed, and it also contains dust, so that optical light cannot penetrate it.

> **The Interstellar Medium (ISM)**
>
> The space between stars is not empty. It contains gas and dust. This is called the interstellar medium, or ISM.

Charged (ionized) gas

Ionized gas is found around young, massive stars that have very strong ultraviolet radiation fields around them, which can separate the proton from the electron in a hydrogen atom, thus creating a cloud of charged protons and electrons. The JWST picture (next page) of Ophiuchus shows ionized gas. This is the bright white and yellow material in the lower centre of the image. The newly formed bright star can be seen in the middle of the bright yellow emission. The star is bright because it is a young high-mass star. Some of the yellow and orange emission is also light from the bright star being reflected and scattered by the dust in the cloud.

The cycle

All of these types of gas are present in the ISM of any disc galaxy at different times in different places. In fact the gas can be cycled between the different types depending on what is happening in its surroundings. The whole of the ISM in every spiral galaxy is one giant recycling plant, churning gas between its different types and building new stars along the way. Atomic gas can turn into molecular gas if the density is high enough relative to the rest of the medium – although it is still close to a vacuum. When this happens, there is sufficient dust to keep out the ultraviolet light from the galaxy's radiation field, and so the temperature becomes very low, sometimes only 5 to 10 degrees above absolute zero, or around minus 263 to minus 268 Celsius. In these conditions, the gas freezes onto the dust grains, and molecules are made on the surfaces of the grains.

A picture taken in the near infrared by the James Webb Space Telescope of a region in the constellation Ophiuchus at a distance of about 400 light-years within the plane of our Galaxy.

The dense regions are where stars are formed, so if a star forms in the cloud, then the star's radiation heats everything again. When this happens the molecules evaporate from the surfaces of the dust grains, and we have a molecular gas. If the gas remains in the vicinity of the hot star and the star is hot enough, then the molecules will come apart (dissociate) in the star's radiation field and return to being atoms.

Labelling the Three Phases: HI, HII and H_2

The three types, or phases, of hydrogen are given different symbols, which originated in laboratories studying gases. Atomic hydrogen is labelled HI to signify the neutral atomic phase. Charged, or ionized, hydrogen is labelled HII. Confusingly the Roman numeral II means it has lost only one electron. Regions containing large amounts of ionized hydrogen are called HII regions. Furthermore, by continuation, HeIII is the charged result after a helium atom has lost both of its electrons, and OIV is an oxygen atom that has lost just three electrons to become an ion (an ion is a charged atom). So the number of lost electrons is always one less than the roman numerals. Finally, a hydrogen molecule, which consists of two hydrogen atoms, is given the symbol H_2 to signify the two atoms. Similarly, for example, a water molecule is H_2O, for its two hydrogen atoms and one oxygen atom.

A Mosaic of Part of the Milky Way

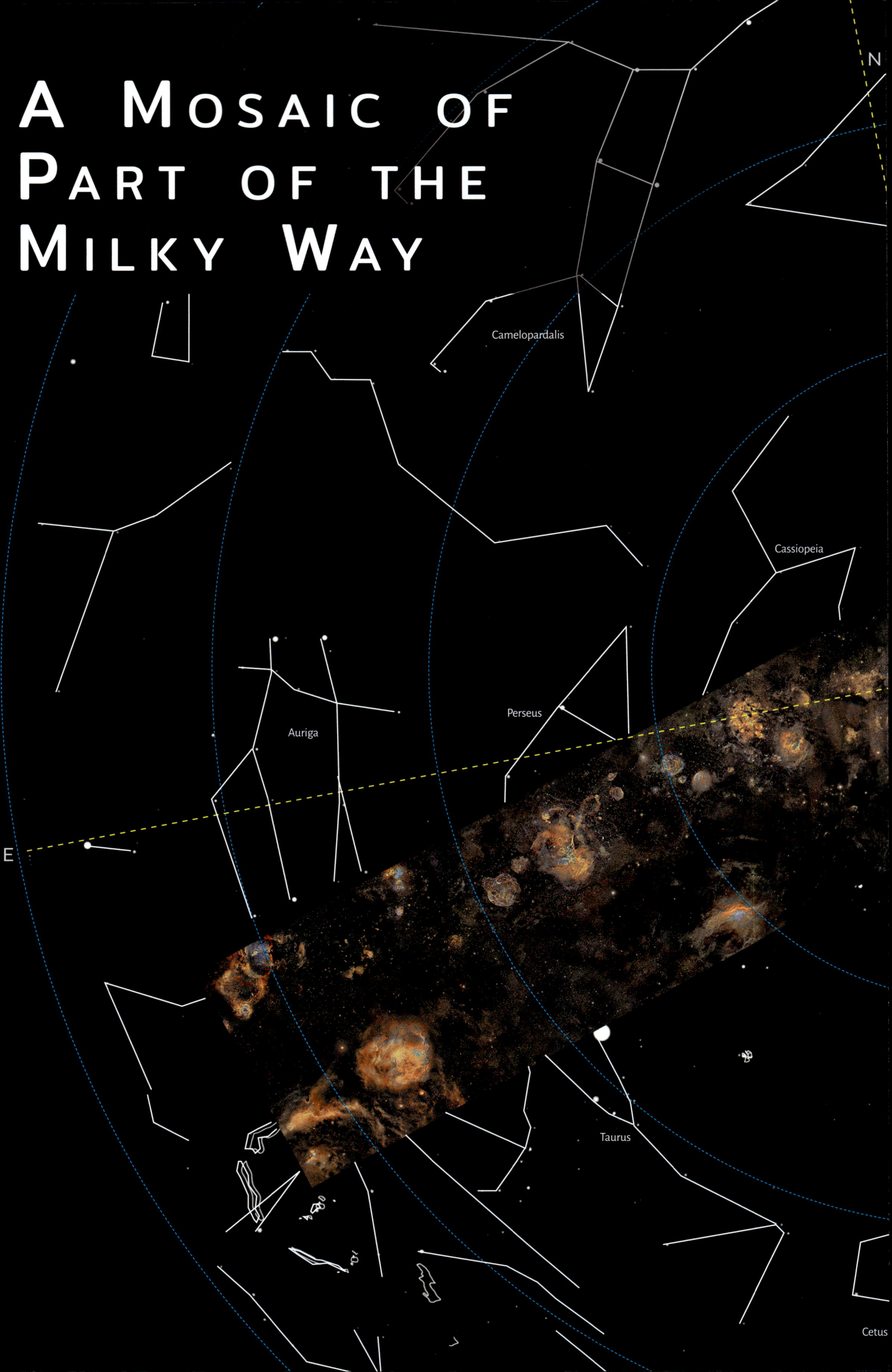

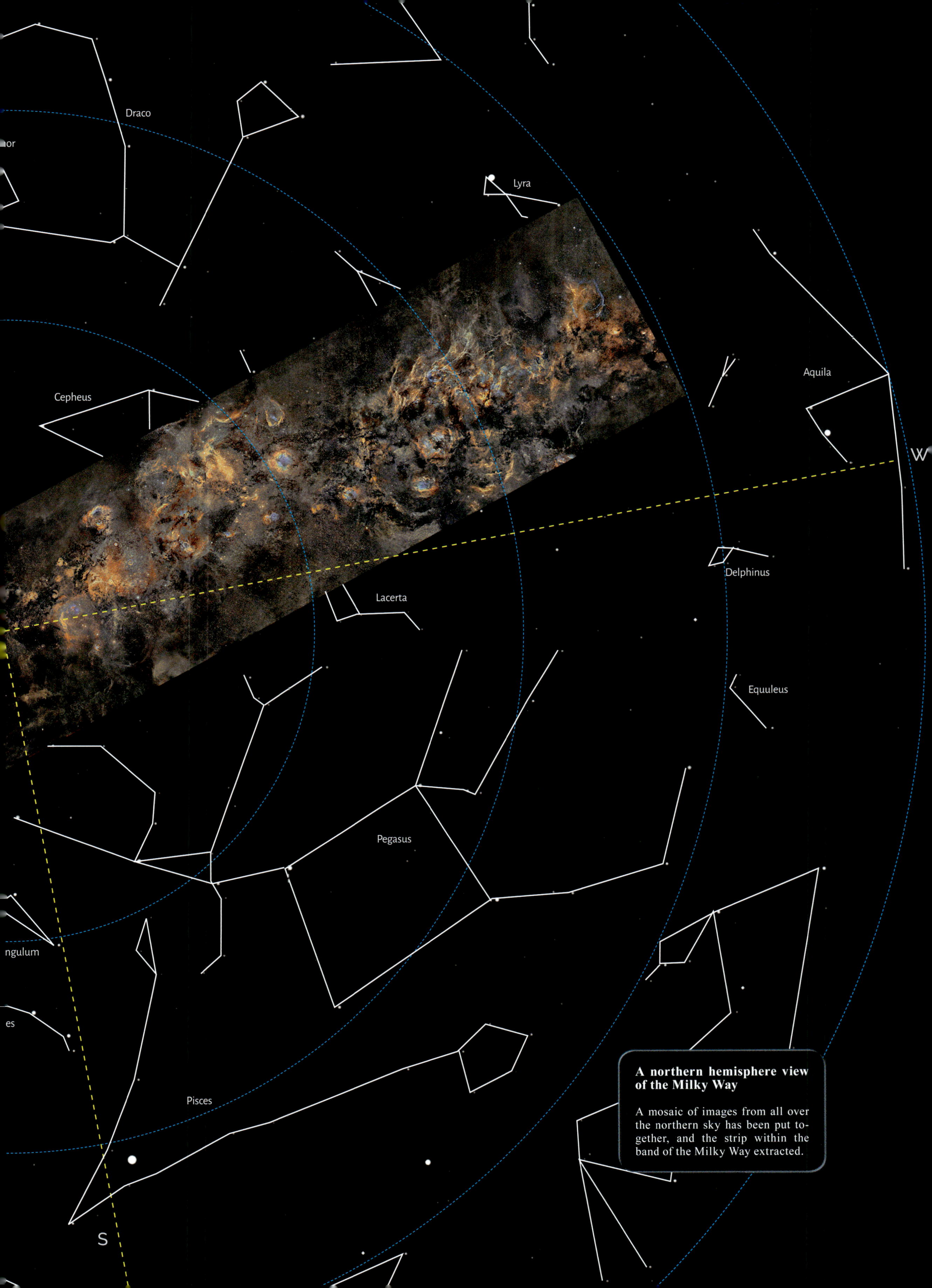

A northern hemisphere view of the Milky Way

A mosaic of images from all over the northern sky has been put together, and the strip within the band of the Milky Way extracted.

A Mosaic of Part of the Milky Way

One of us (J.-P. Metsävainio) is not only an avid photographer and collector of astronomical images, but has also developed new working and imaging methods to make use of the maximum amount of data possible. He has produced the most detailed image of the largest stretch of the Milky Way Galaxy ever published. This has taken over a decade of work, on and off, and we present the results here. This is a very brief description of how it was done.

Astronomical photography is a highly time-consuming process. To produce a colour image, each target had to be shot at least three times through different filters to have the three colour channels needed. Sometimes, the exposure times could be very long – up to hundreds of hours for some very dim objects. An average exposure time was around 25 hours, and everything had to be carefully planned.

The images were taken using many telescopes over more than a decade until the mosaic was completed. Then the mosaic had to be put together. This is not as easy as tiling a floor, because what we actually see is the sky apparently

J.-P. Metsävainio's giant stereoscopic mosaic of the portion of the Milky Way visible from Earth's northern hemisphere. To view this in stereo, please rotate the book 90 degrees anticlockwise.

projected onto the inside of a sphere – similar to a geographical map on the outside of a globe of Earth. Consequently, the sky is effectively distorted relative to a flat page in a book, and this distortion must be taken into account – rather like a map projection of Earth in an atlas. This required special software to ensure that the projection was accurate. We hope that you agree that the finished product was worth the work. The mosaic allows us to study our Galaxy in great detail, but to understand what we are looking at we first need to know about how the Galaxy is constituted.

Details of the Mosaic

The Milky Way is like all galaxies introduced in Chapter 2, in that it consists of dark matter, normal baryonic matter, a radiation field that is the sum of the radiation emitted by all the stars, and a magnetic field that is caused by the charged particles moving throughout the Galaxy. For now we will concentrate on the normal baryonic matter, since that is what is shown in our photographs; we will consider the other constituents later. Baryonic matter is what makes up the stars, the planets, us and everything we can see.

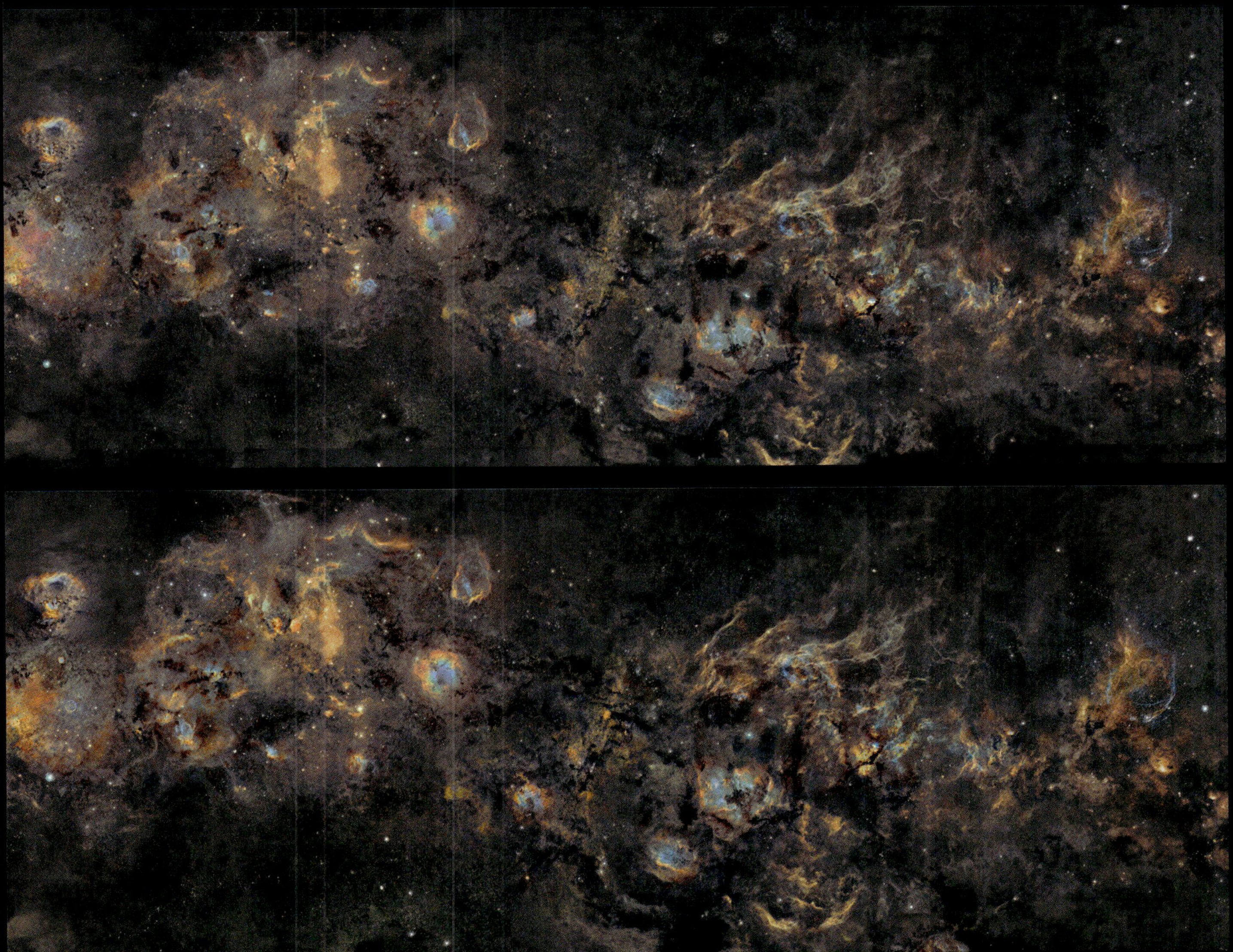

IC 410

IC 410 is a large star-forming region in Auriga at a distance of about 12,000 light-years. It contains atomic gas around its edges, which has become molecular gas in its denser regions. Mixed in with the gas is dust, which we see in a variety of colours as it scatters light at different wavelengths, depending on the various sizes of the dust grains. Some of the colouring is also due to the different atoms, molecules and ions emitting at different characteristic wavelengths. Some stars have already formed in this region, but many more are still forming.

The star-forming region IC 410 in Auriga. The star formation has burst out of the near side of the cloud.

The Sharpless Catalogue (Sh)

The Sharpless Catalogue of ionized regions was compiled by Stewart Sharpless (1926–2013) in 1959. He selected regions of ionized (charged) gas that are found around newly formed high-mass energetic stars. Such regions are called HII regions. Note that the Sharpless Catalogue does not contain any galaxies.

Sharpless 2-232

Sharpless 2-232 is another region where stars are actively forming, and some massive stars have already formed. Consequently, it contains some atomic gas, some ionized (charged) gas and some molecular gas. In the centre of this roughly rounded object there is an HII region. This is obscured by the region of molecular gas and dust seen in yellow and blue, with atomic gas around the edges. It lies at a distance of over 5,000 light-years. There are some massive young stars in the middle, illuminating the region and ionizing some of the gas.

IC 1848

IC 1848 is a further region of star formation, somewhat similar to Sharpless 232. The blue regions represent where the hotter gas lies, which is towards the centre, and the yellow, orange and reddish regions trace the cooler gas and dust on the outside. Hence, we note that we can see into the central regions of this object and the cooler, outer regions have been blown away on the side nearest us. This probably means that the area where the hottest regions lie has burst out of the edge of this object, which is forming stars in the nearer half of the cloud, and so we can see into the middle. It lies at a distance of about 7,500 light-years.

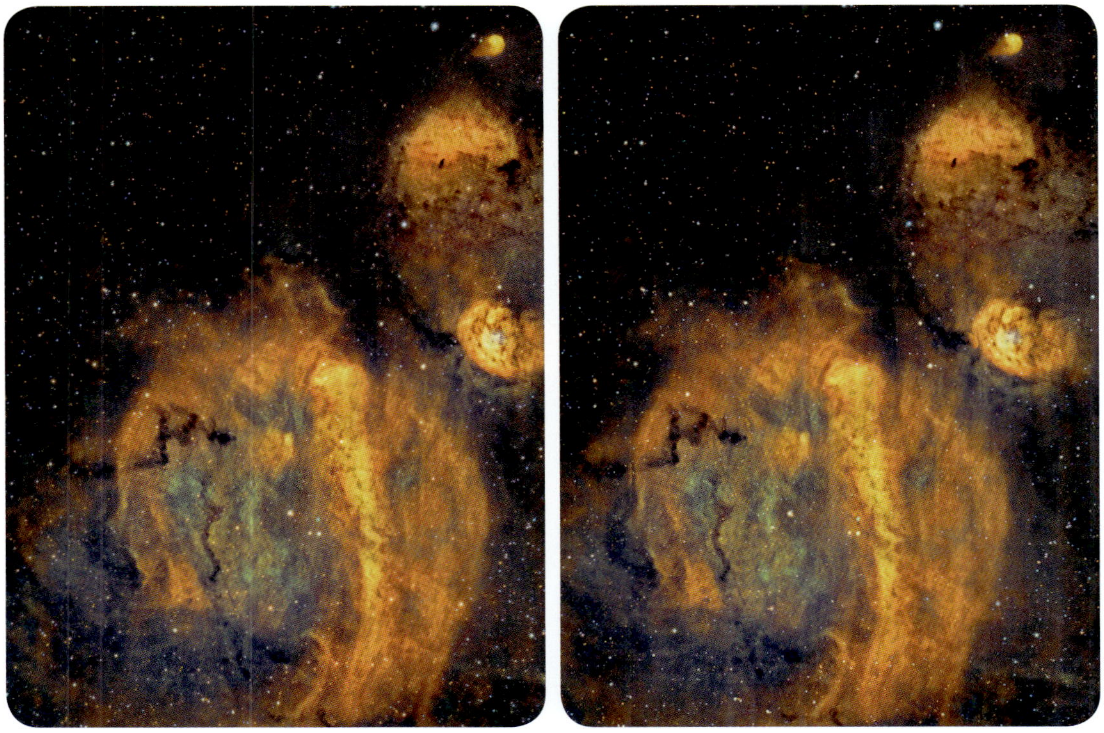

The star-forming region Sharpless 2-232 in Auriga. Golden areas are from emission of sulphur and hydrogen. There is also a little ionized oxygen, OIII, seen as the blue-green regions.

The star-forming region IC 1848 in Cassiopeia. It can be seen that this is a more active region, forming high-mass stars, because there is more ionized oxygen, OIII, which requires energy from high-mass stars to produce it.

Planetary Nebulae

The term 'planetary nebula' is a complete misnomer, as planetary nebulae have nothing to do with planets at all. They were originally thought to be signatures of newly forming planets. It is now realized that they are much too large for planetary systems. They are known to form around evolved stars, which jettison their outer layers into space as they pass through their later evolutionary stages. These outer layers are seen as partially transparent structures as they continue into space, often forming interesting shapes.

Outters 4

This beautiful spindle-shaped object is the planetary nebula named Outters 4 after its discovery by Nicolas Outters in 2011. It is believed to lie at a distance of over 2,000 light-years, and to be driven by a triple system of three massive stars in its centre. Usually, planetary nebulae are roughly round, but this one clearly is not. It is believed that there must have been an outflow from the poles of the star system, which created the elongated structure, followed by the more spherical, normal jettisoning of the outer layers of the evolved stars.

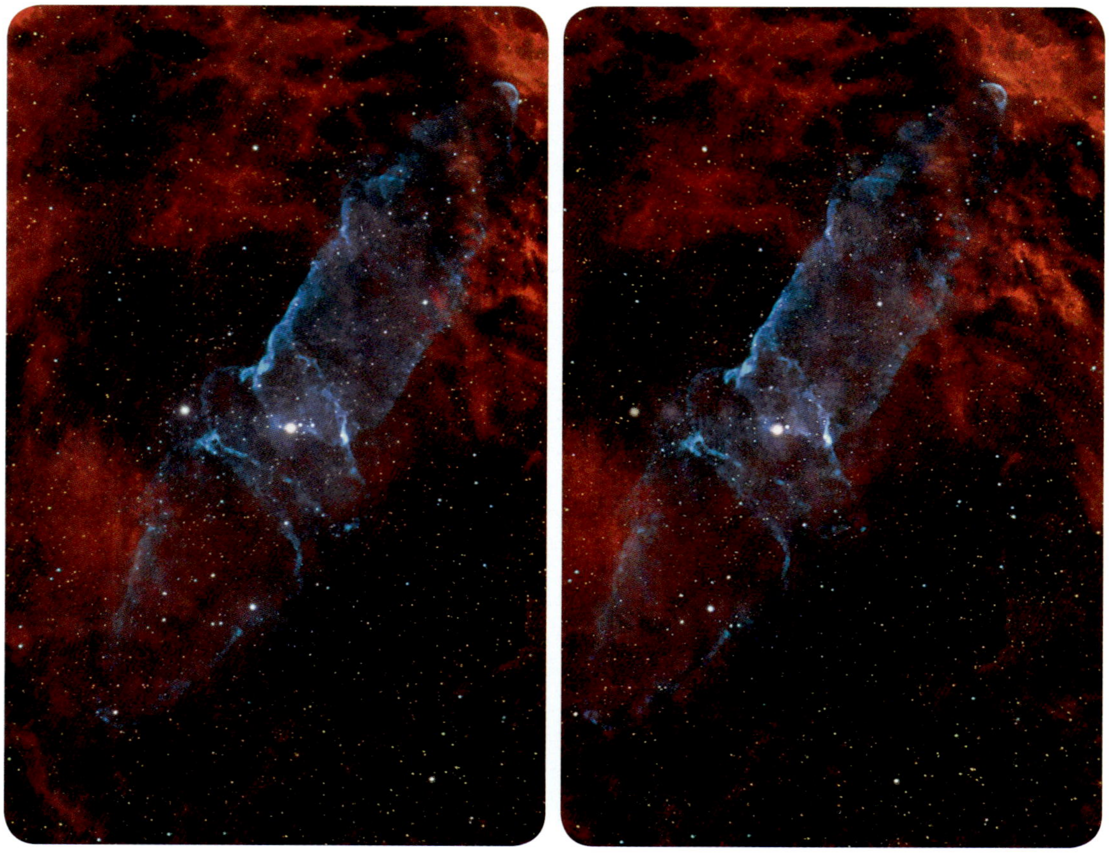

The planetary nebula Outters 4 in Cepheus. Note the unusual cylindrical shape of the nebula, probably caused by outflowing gas from the poles of the central star system.

Sharpless 2-188

Sharpless 2-188 is a more typical planetary nebula than Outters 4, with a clearly rounded shape. The image shows a shell-like structure surrounding a central star. This shell is the remnants of a shell of matter that was jettisoned from this evolved star during the latest period of activity. The shell is expanding into space and will eventually become so thin and indistinct that it will no longer be visible. It is not fully symmetric, with one side being brighter than the other, possibly because the star is moving with respect to the local interstellar medium, making the leading edge brighter. Alternatively, some asymmetry in the close environment of the star may have caused this asymmetry in the shell.

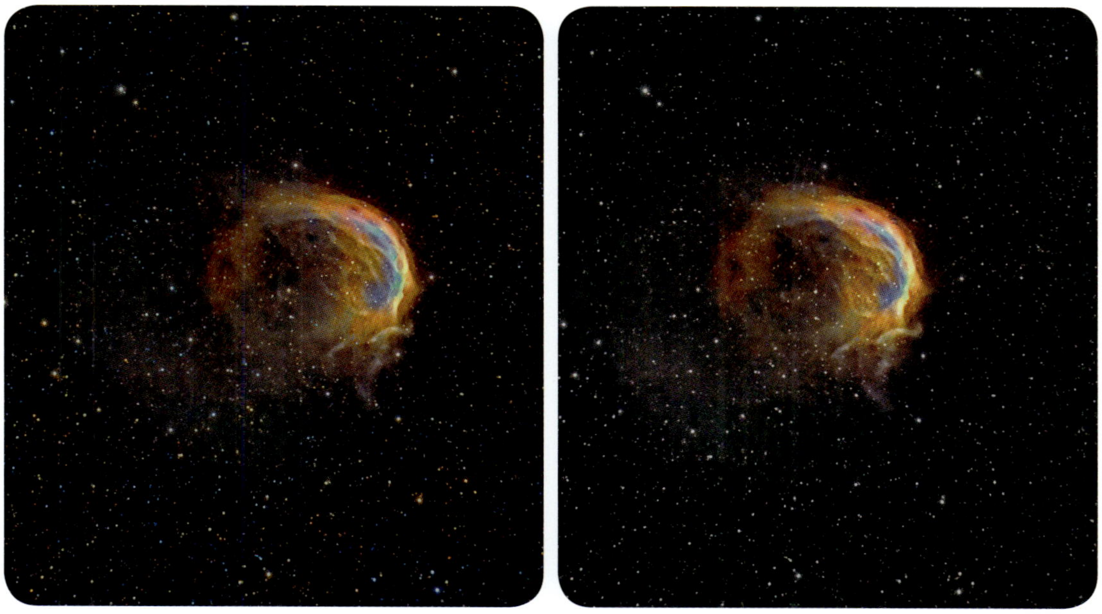

The planetary nebula Sharpless 2-188 in Cassiopeia. The more spherical shape of this nebula indicates that it was caused by a more symmetrical expansion.

Sharpless 2-224

Sharpless 2-224 is a supernova remnant (SNR) in Auriga, at a distance of just under 15,000 light-years, which exhibits an unusual shape. In one half it shows the classic hemisphere shape that would be expected of the symmetrical expansion of a shell around an exploding star.

However, the other half of the sphere is missing, and instead, further out, we see a conical shape (upper left). Astronomers postulate that the asymmetry may be caused by an asymmetry in the environment of the SNR whereby on one side the interstellar medium has its usual density, but on the other side it is much more sparse to begin with, and then there is a sudden increase in density. The usual density side generates the usual hemispherical result (lower right), but on the other side the expansion carries on unrestricted until it hits higher density matter. The expansion wave is what we call a 'shock wave' because it moves at, or close to, the speed of sound, much like the shockwave generated by a supersonic aircraft.

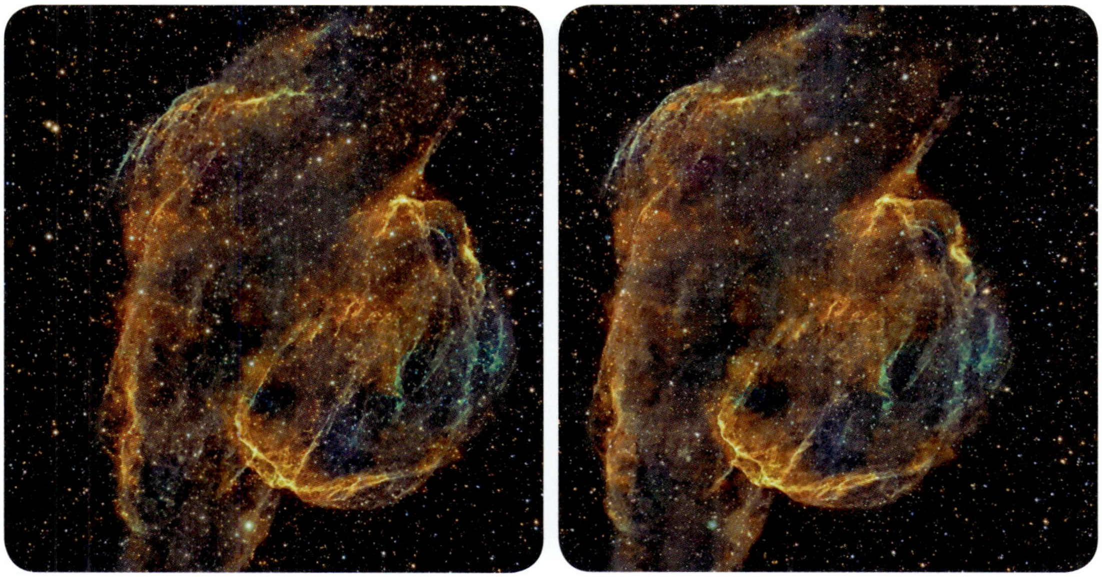

The supernova remnant Sharpless 2-224 in Aurig. Note the cone shape of the shock in the upper left of the image.

> **Supernova Remnant (SNR)**
>
> A supernova remnant (SNR) is created after a star undergoes an extremely violent explosive event at the end of its evolution. There are two types of supernova explosion. One occurs after a very massive star runs out of fuel to burn and ends its evolution explosively before collapsing to a black hole. In the other, a binary star consisting of a white dwarf star and a red giant behave in what is known as a 'symbiotic' manner: the red giant dumps so much matter onto the surface of the white dwarf that the smaller star exceeds its critical mass and explodes. The matter blasted into space by any form of supernova expands in all directions, creating a shell of matter drifting outwards through space.

Supernova remnant CTB 1

CTB 1 is a more usual round supernova remnant, with a shape that looks vaguely reminiscent of a mushroom cloud. It is quite easy - especially in stereo - to see this SNR as an explosion from a dying star. Its designation of CTB 1 comes from its listing as number 1 in *Caltech Observatory List B* (CTB), which is a catalogue of radio sources published in 1960.

One of the possible end-states of a star, after it has exploded as a supernova, is to become a pulsar - a rotating neutron star. It has been found that CTB 1 does, in fact, host a pulsar known as PSRJ0002+6216. The SNR is about 9,600 light-years away.

All the types of objects discussed in the section above inhabit the interstellar medium of our Galaxy.

The supernova remnant CTB 1 in Cassiopeia. Again the spherical nature of the nebula indicates a symmetrical explosion.

Outside the Plane of the Galaxy

If we look above and below the plane of the Milky Way Galaxy, we see what we refer to as the 'halo'. This is where we find the dark matter that we discussed earlier. However, this is also where globular clusters are seen.

Globular clusters are clusters of stars that are packed much more closely together than in the plane of the Galaxy. They are also typically older stars than those found in the plane. It is believed that globular clusters formed first, before the rest of the Galaxy started forming stars, and while the Galaxy was still collapsing onto the plane.

Globular cluster M13/NGC 6205

M13, or NGC 6205, is a typical globular cluster in the halo of the Milky Way. It lies at a distance of just over 20,000 light-years in the constellation Hercules. It is believed to contain up to half a million stars and is 100 to 200 light-years across. This is just one of over a hundred clusters in the halo of the Milky Way. All large galaxies seem to host globular clusters in their halos.

The globular cluster M13. A globular cluster is an agglomeration of stars that all formed at about the same time in roughly the same place, usually outside the plane of a galaxy. They are typically among the oldest stars in a galaxy.

The Inner Galaxy

The dynamic centre of our Galaxy, its centre of mass (called the barycentre), appears to coincide with the object that astronomers call Sagittarius A* (Sgr A*), which is now known to be a black hole. Let us take an imaginary journey through the inner parts of our Galaxy on our way to Sgr A*.

The Milky Way has been classified a number of times, but the current consensus among astronomers is that it is probably of type SBbc. This denotes that it is a barred spiral galaxy (SB) with intermediate to open, clearly defined, spiral-armed structure (bc). See Chapter 4 for further discussion of galaxy types.

The fact that the Milky Way is a barred spiral and not just a spiral was only realized in the late 1980s and early 1990s, and during this time there was a debate on the matter. Infrared images from the NASA Cosmic Background Explorer (COBE) satellite showed a distinct asymmetry, indicating that we are looking at the bar just slightly to one side of end on, and the brighter side is closer to us.

The Bar and Bulge

Our Sun sits at a distance of about 25,000 light-years from the black hole Sgr A*. As can be seen from the rendering on the next page, which is based on counting stars at infrared wavelengths, the inner part of the Galaxy consists of an oval-shaped distribution of stars and an extended feature that the astronomers call a 'bar'. The oval of stars can be seen in photographs taken from Earth, but the bar cannot usually be observed in photographs because we view it close to end on. This bar could extend to 15,000 light-years from Sgr A*.

The oval-shaped feature is what astronomers call the 'bulge' of our Galaxy. It is also oval in cross-section, and it extends above and below the plane of the Galaxy, so it is visible from Earth. In fact, on a clear night from a dark sky, it can be seen with the naked eye. The elongation of the bar is aligned with the elongation of the bulge. Astronomers estimate that the long axis of the bulge could be 8,000 light-years in radius.

In recent years it has been realized that galaxy bulges and bars are more complicated than mere ovals. This is particularly true in our Galaxy. Astronomers have used the method of counting stars to estimate the local density in different parts of the bar and work out a three-dimensional shape for it. The infrared is used to see more deeply into the centre. Then it is possible to deduce what the bar looks like from different angles.

It can be seen (next page bottom) that the bar is not in fact oval shaped. The shape that is seen has been variously labelled by astronomers as either 'peanut-shaped', because of a narrowing in the middle with a broadening either side, or 'box-shaped', owing to it being closer to a rectangle than an oval. There are seen to be extensions towards each of the four corners of the image, and these have led some to refer to it as 'X-shaped'.

The bulge and bar are not static, however, but dynamic, and they certainly do not behave anything like a single structure. As telescopes have pushed further and further into the inner Galaxy, they have shown more and more structure.

A genuine (not an artist's impression) infrared picture of the inner region of our Galaxy, taken by the NASA COBE satellite. The inner part of the disc can be seen. The extension above and below the disc is actually the bar in the centre of our Galaxy. Notice how the right-hand side of the bar appears brighter (and hence denser) than the left-hand side. This leads astronomers to believe that we are looking at the bar not quite end on.

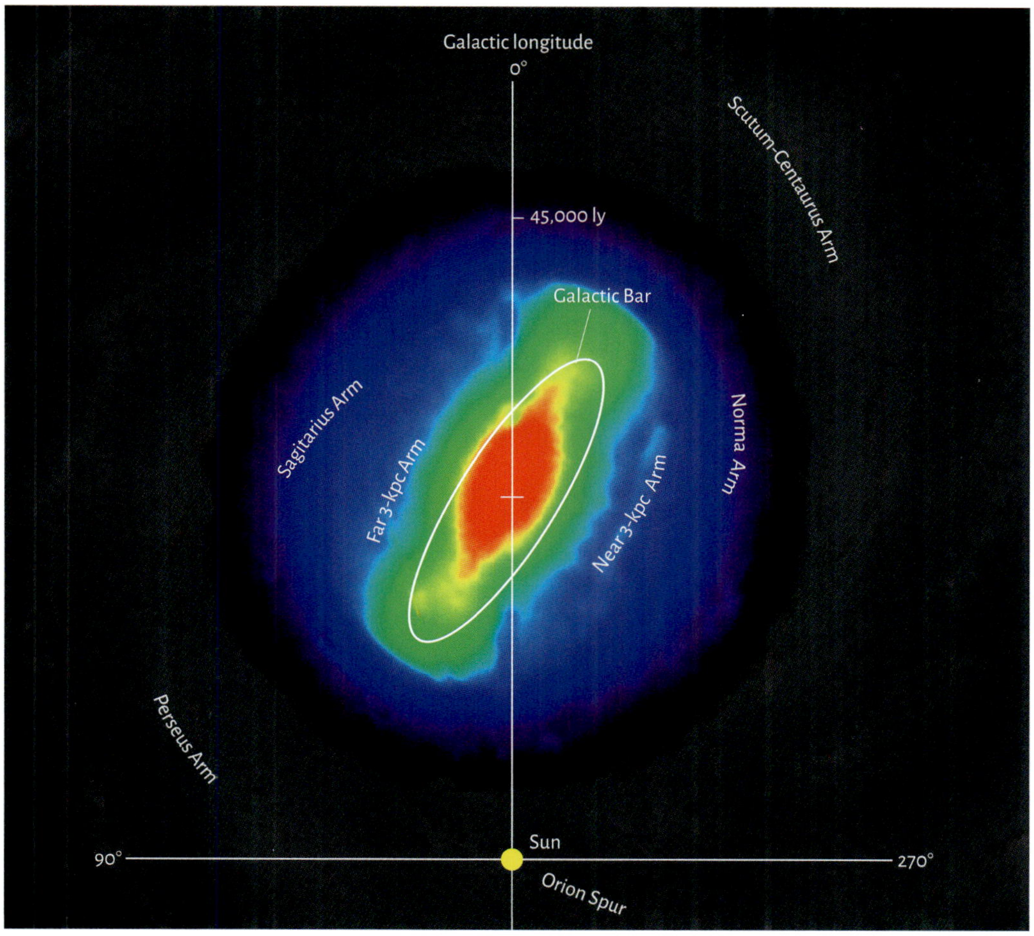

A suggested rendering of the central regions of our Milky Way Galaxy seen from 'above' (perpendicular to the plane of the Galaxy), showing the bar and other features.

Why do Astronomers Use the Infrared?

Firefighters in a smoke-filled building will sometimes use infrared cameras to help them see anyone trapped in the building. Astronomers have a similar problem seeing through the dark lanes of the Milky Way. Infrared cameras in astronomy perform the same job of seeing through these dark lanes and other dust clouds.

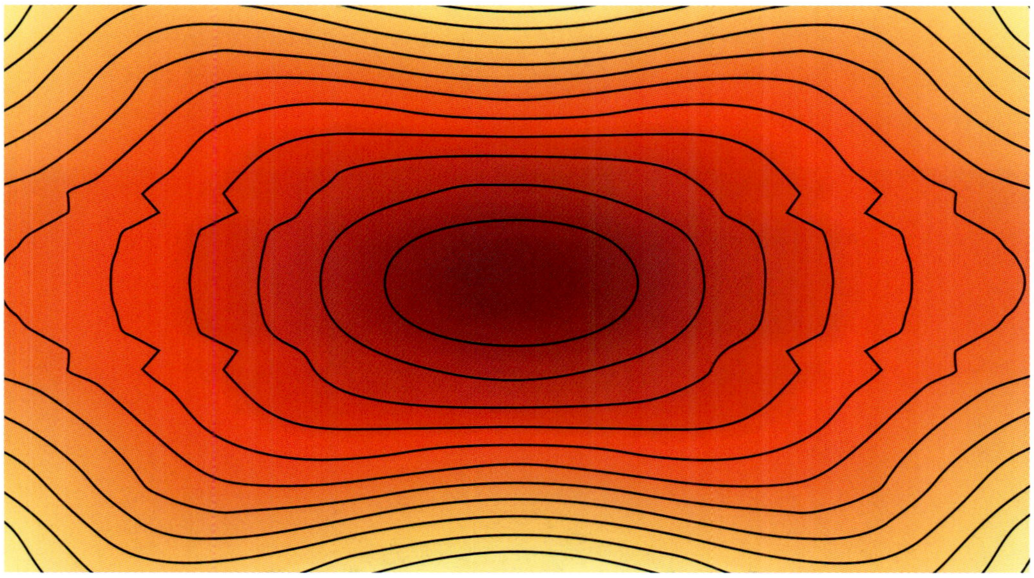

An estimated view of the bar of the Milky Way Galaxy if we could see it side on rather than the almost end-on view that we are presented with. The shape of the contour lines of equal density show why astronomers have variously designated it 'peanut-shaped', or 'box-shaped' or even 'X-shaped'.

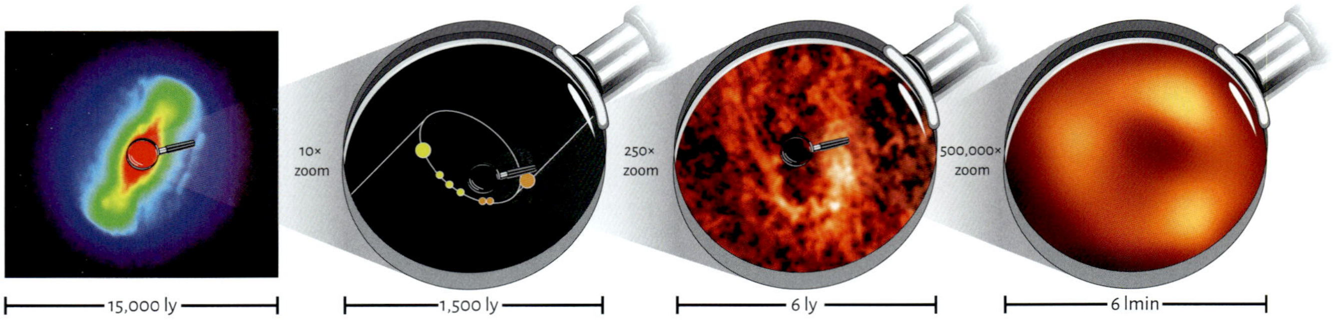

Progressive zooms in on the inner Galaxy, shown in detail on the next page.

Journey to the Centre of the Galaxy

The inner ring or central molecular zone

The diagrams (a) to (d) on the next page zoom in by progressively larger steps in magnification until we reach the black hole itself. The first image, (a), shows the bar and bulge that have already been discussed. The second image, (b), is zoomed in by a factor of ten, and suddenly a structure has appeared that looks like a smaller-scale version of another spiral galaxy. The spirals are dust lanes of matter streaming in towards the inner Galaxy.

Both of these images are predictions of what we would see if we were looking 'down' on our Galaxy, at right angles to our view from Earth. The second image is actually based on computer models of the Galaxy using all the available information and data. The inner ellipse seen in the image is matter orbiting the Galactic Centre, which is formed of clouds of molecular gas, so is often referred to as the 'central molecular zone' (CMZ).

The circumnuclear disc and inner cavity

The third image, (c), zooms in by about another factor of 250 and shows what astronomers call the 'circumnuclear disc' (CND). This is a disc of matter that surrounds the innermost region of the Galaxy and is only a few light-years across. It can be seen clearly in this image of radio emission from the molecules in the gas in the disc, which was taken by the Atacama Large Millimetre Array Radio Telescope in Chile. This is a real image rather than an illustration, and it shows that the disc is actually rather uneven and not exactly symmetrical.

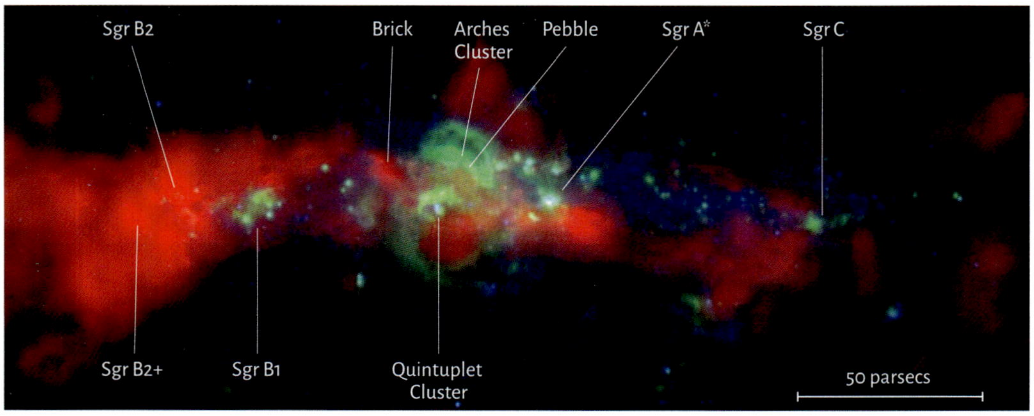

A real (but false-colour) set of images of the structure of the central molecular zone modelled in image (b) on the next page as seen from Earth, where we see the structure edge on. The various clouds have their astronomical names (Sgr is short for Sagittarius). Each of the colours, red, green and blue, represents the emission from different wavelengths in the clouds, and thus a different type of gas.

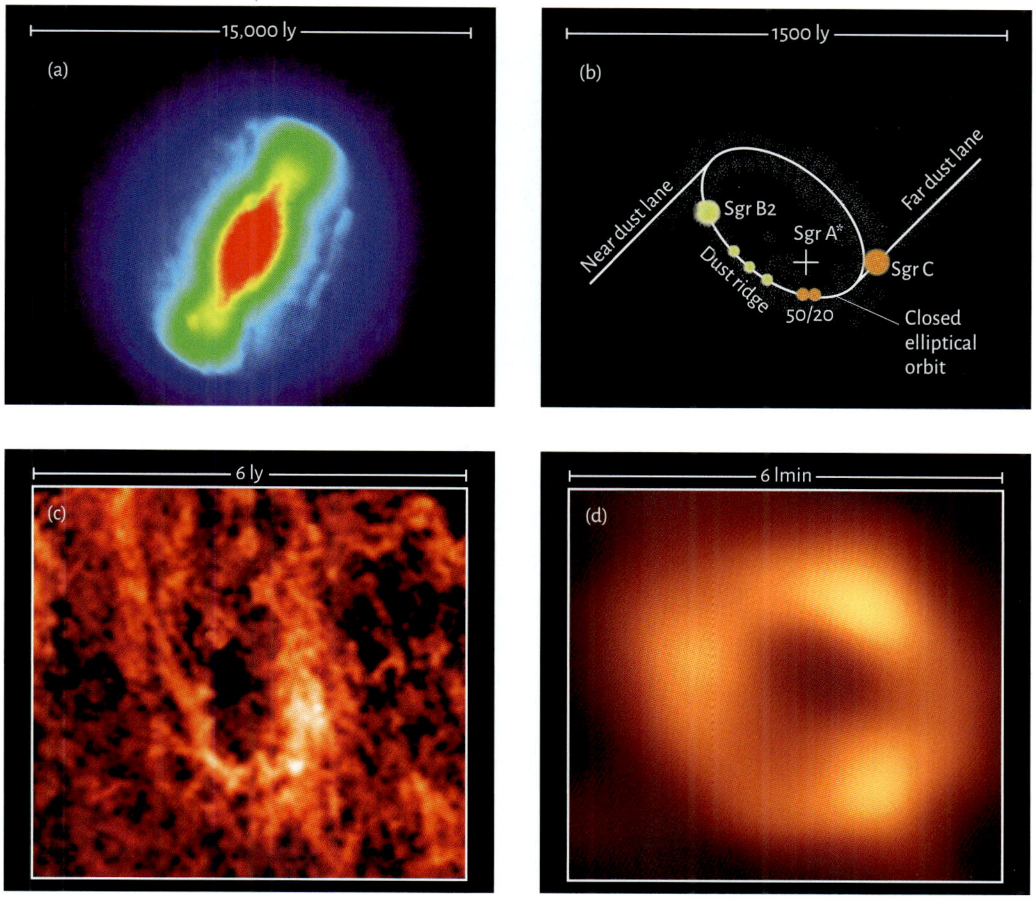

Top two: artist's impressions of the Galaxy seen from 'above': (a) the bar of the Galaxy; (b) the central molecular zone at the centre of the bar. Bottom two: actual pictures of (c) the Galaxy's circumnuclear disc; (d) the central black hole, Sgr A*, and its surrounding disc.

In the gap that can be seen in the centre of the ring there is much less matter than in the ring itself. This gap in the centre is known as the 'central cavity', although that is crucially where the black hole is – on a very much smaller scale again.

The central cavity was referred to in Chapter 1 when we discussed why the Milky Way is no longer an active galaxy. The general paucity of matter in the central cavity means that there is relatively little matter available to fall into the black hole. Matter falling into a black hole is what is believed to power the jets emitted from the surroundings of black holes, which in turn define whether or not a galaxy is active.

The black hole at the centre of the Milky Way

Our hypothetical journey is now approaching its end. The fourth picture (d) is at a scale 500,000 times smaller than the third picture (c) and shows Sgr A*, the black hole at the centre of our Galaxy. To make such an enormous leap in scales visible to a telescope, astronomers needed a enormous leap in telescope technology. This was provided by the Event Horizon Telescope, which required a series of technological breakthroughs that had been previously thought to be in the realms of science fiction – and are described on the next page.

When the image of the M87* black hole (see Chapter 2) was released in 2019, it caught the attention of the global media and was shown on almost every media outlet in the world. However, of more interest to Milky Way astronomers was the image of Sgr A* above in image (d) that was subsequently released.

The Event Horizon Telescope

We will digress slightly to look at how the technological breakthrough of the Event Horizon Telescope (EHT) came about. The EHT is not a single telescope but is rather a number of radio telescopes around the world, linked together. Astronomers had linked radio telescopes before in a process known as interferometry. Radio telescopes had even been linked around the world before, but only at centimetre wavelengths.

The new aspect of the EHT was the fact that it worked at millimetre rather than centimetre wavelengths. Millimetre wavelengths are what microwave ovens use. A major consequence of working with wavelengths ten times smaller is that all the tolerances on all of the equipment have to be ten times tighter. This is extremely challenging technologically.

However, with wavelengths ten times shorter, this telescope array can see things that are ten times smaller, as discussed in the box on resolution in Chapter 2. The resolution of the EHT is about 20 millionths of a second of arc, where 1 second of arc is 1/3,600 of a degree. It is hard to visualize an angle this small, but as a guide, if an optical telescope on Earth achieved a similar resolution, it would be able to photograph a tennis ball on the Moon!

The EHT Telescopes

The EHT telescopes all operate at microwa wavelengths of about a millimetre, the high fr quency end of the radio spectrum. As they co lect signals simultaneously around the world, th data must be captured in such a way that it can b time-stamped to then be put together afterward At the speed of light the waves travel a millimet in roughly three million millionths of a secon Only atomic clocks can measure this accuratel so every telescope has a perfectly synchroniz atomic clock. When all of the signals are com bined they can be made to seem to have com from one big telescope the size of Earth. Th requires a large amount of time on the bigge supercomputers to make a picture. However, f the signals to appear to come from one giant tel scope, the astronomers must know the positions each of their telescopes to the nearest millimetr This is also extremely challenging but can be do using laser ranging and satellites.

Achieving such resolution was most definitely in the realms of science fiction only a few short years ago. But now, not one but two black holes have been imaged, M87* and Sgr A*. The fact that they both appear to be roughly the same size is a pure coincidence. M87* just happens to be 1,800 times further away and also 1,800 times bigger in linear extent than Sgr A*. This was the coincidence that the EHT astronomers took advantage of in order to take pictures of both.

The galaxy NGC 4565, which appears most like our Milky Way seen edge on externally. First observed in 1785 by Sir William Herschel, this is one of the most famous examples of an edge-on spiral galaxy and is located some 30 million light-years away in the constellation Coma Berenices.

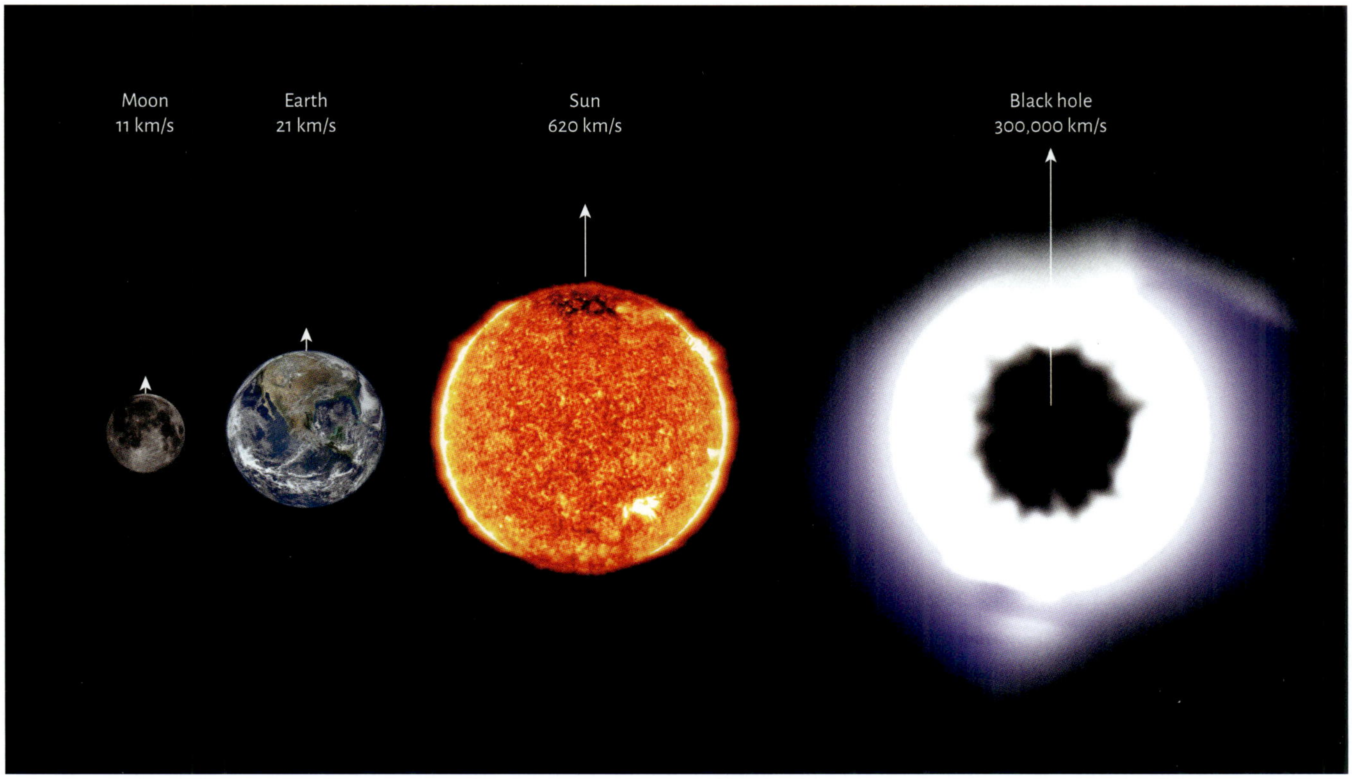

The escape velocity of four different bodies of increasing mass.

The first pictures of black holes

The picture of Sgr A* taken by the Event Horizon Telescope is interesting both in its similarities and differences to the picture of M87* seen in Chapter 2. Both images have a central dark region coincident with the black hole event horizon, surrounded by a brighter ring.

The ring is a combination of light from the disc of matter surrounding the black hole event horizon and from behind the black hole. We see the light from behind because of the effect of general relativity, whereby a large mass like a black hole bends space, causing light to appear to bend around the black hole. Astronomers are unsure as to the nature of the bright spots, but they may simply be slightly denser regions in the so-called 'accretion disc' around the black hole. Matter is flowing in (accreting) through this disc onto the black hole itself.

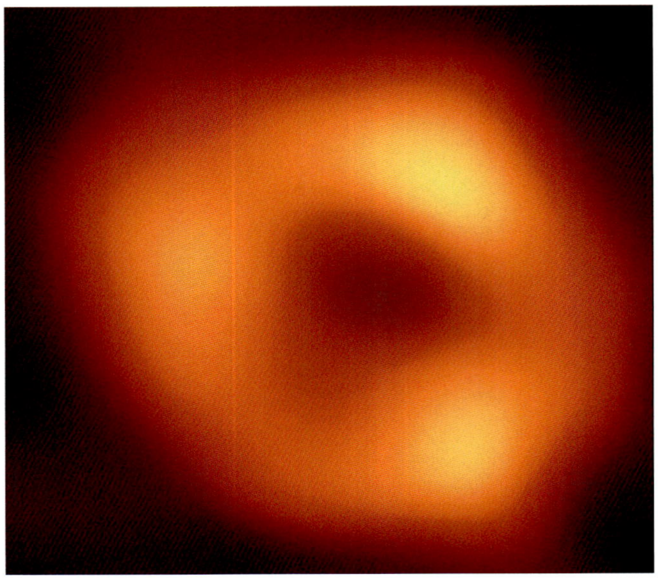

The picture of Sgr A* taken by the Event Horizon Telescope.

Escape Velocity

Every astronomical body has an 'escape velocity'. This is the speed that you need to be travelling in order to escape from the gravitational attraction of the object. For instance, for a rocket to leave Earth it must be travelling upwards at a speed of 11 kilometres per second. The more massive the object, the greater the escape velocity, so to escape from the surface of the Sun a rocket would need to be travelling at over 620 kilometres per second.

Albert Einstein's theory of special relativity, which says that nothing can travel faster than the speed of light, predicts that there can be objects from which nothing can escape. His theory of general relativity predicts how a black hole could curve space itself.

The picture of the black hole is only 6 light-minutes across (the distance travelled by light in 6 minutes, or over 100 million km), which is less than the distance from Earth to the Sun. The inner edge of the disc is less than that distance, or only about 30 million kilometres in radius.

Every black hole is predicted to be surrounded by a so-called 'event horizon'. This is the point at which the escape velocity from the black hole becomes greater than the speed of light. Einstein's theory of special relativity tells us that nothing can go faster than the speed of light. Therefore nothing can escape from inside the event horizon of a black hole.

Astronomers can model what they believe is happening in the disc surrounding the black hole. This requires the use of Einstein's theory of general relativity, which states that space is warped by massive bodies such as black holes. It is difficult for us to imagine this, but the best analogy is a two-dimensional surface, curved into a third dimension, as in our artist's impression.

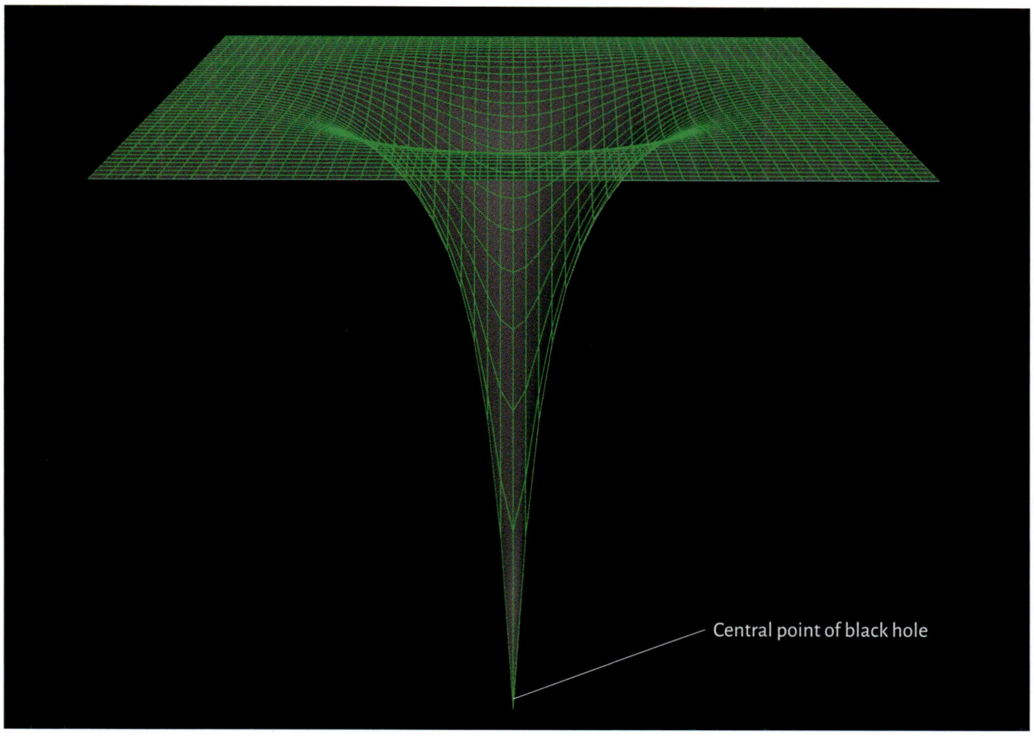

The gravity of a black hole: an artist's impression of a black hole curving space.

In the image, we are imagining the three dimensions of space displayed simply as two dimensicns, such as with a tightly stretched rubber sheet. Now imagine placing a heavy bowling ball on the rubber sheet, which will become distorted. This represents the effect of a massive body in space, such as a black hole. Finally, imagine trying to roll a marble up the rubber sheet from the bowling ball. If the marble is travelling too slowly, then it will never get out of the hole. So if the speed needed to get out of the hole is more than the speed of light, then nothing will ever escape.

This is then as far into the centre of the Milky Way as we can go: the black hole event horizon - the point of no return. In the next chapter, we will begin to explore other galaxies.

The James Clerk Maxwell Telescope (JCMT)

On Mauna Kea, Hawaii, the JCMT is part of the Event Horizon Telescope that captured the first ever image of a black hole. The University of Lancashire (Derek Ward-Thompson) has access to observing time on the JCMT.

Hickson Compact Group 40

An optical image of a group from the Hickson catalogue of compact groups of galaxies, showing both elliptical and spiral galaxies.

Ellipticals, Spirals and Exotics

4

Galaxy Shapes and Types

Edwin Hubble was the first to bring in a classification system of galaxies with his tuning-fork diagram, which we discussed in Chapter 2. As we noted, the diagram does not indicate any form of evolutionary status for a galaxy, but it is nevertheless useful for sorting galaxies according to their appearance. Hubble noted that there are two types of galaxy shapes: elliptical (E) and spiral (S). There is also a third group that does not really fit into either of the other two, which is labelled 'irregular' (Irr).

These classifications have been adapted and modified over the years, with more subgroups and additional subdivisions introduced, but we will stick with the original system, apart from a few minor modifications. For example, galaxies that look mostly like one type of galaxy but have an unusual feature are sometimes designated 'peculiar' (pec).

Elliptical Galaxies

Elliptical galaxies are exactly as their name suggests. Their shapes appear to be ellipses in our pictures of them, and they have no discernible structures within them. They are typically three-dimensional ellipsoids. They generally have no preferred axis along which they are flattened - as spiral galaxies do. Elliptical galaxies are designated by the letter 'E'.

The pictures we take are by nature two-dimensional representations of objects that are three dimensional. So, if an elliptical galaxy has three unequal axes, as is the most general case, then it will appear differently to us depending on the angle from which we are viewing it. Consequently, there is always some ambiguity in our classification of any particular galaxy.

Nevertheless, we can estimate a galaxy's degree of ellipticity by simply measuring its long and short axes and taking the ratio of them. The most symmetrical or spherical are classified E0. An elliptical galaxy with one axis only slightly longer than the other would be classified as E1. The number after the letter 'E' then increases as a galaxy's ellipticity increases.

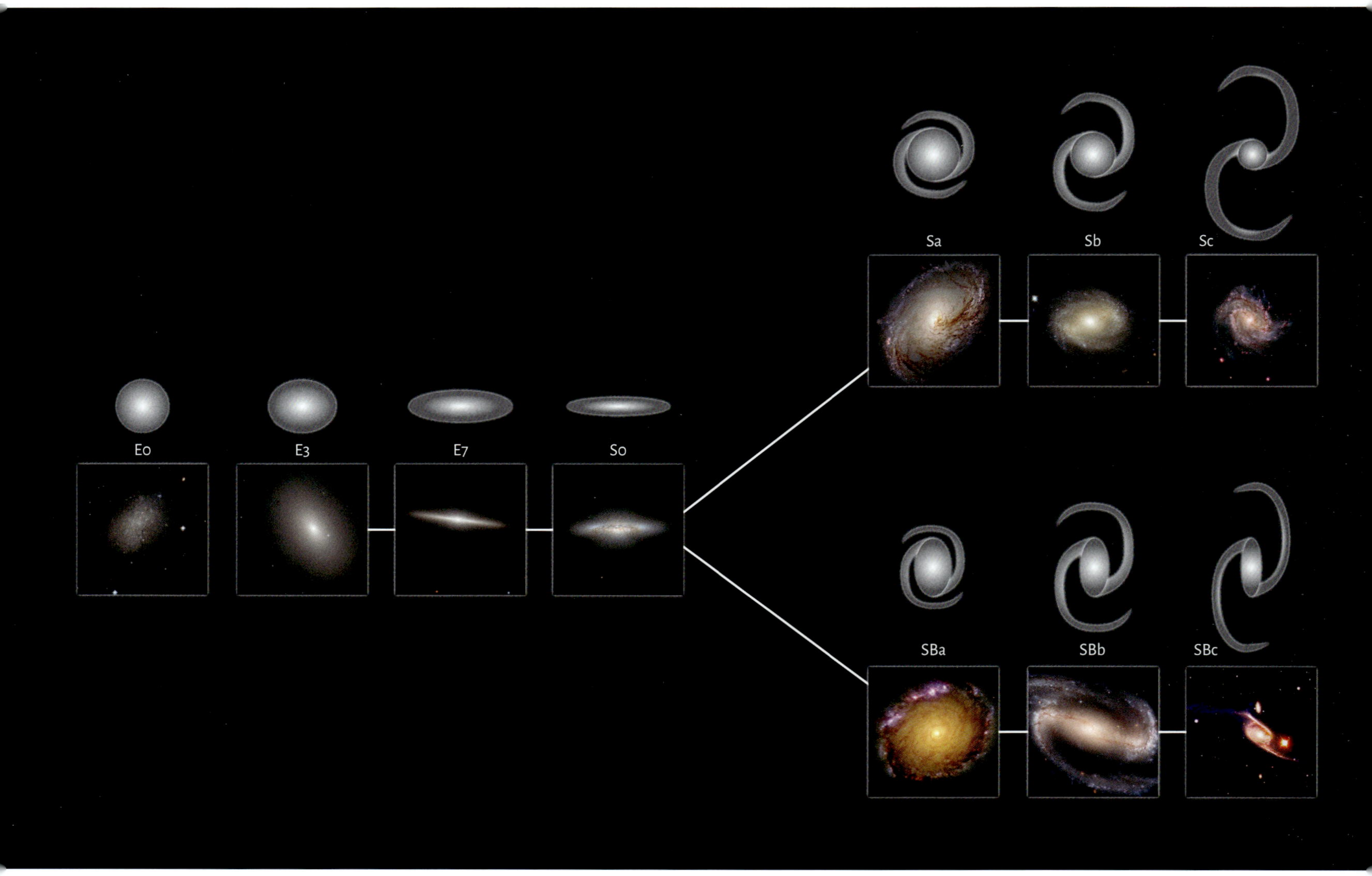

The Hubble tuning-fork diagram of galaxy classification. Elliptical galaxies are at the left, with the two branches of spiral galaxies to the right: normal above and barred below.

When the classification system is drawn, it is usually shown as a tuning-fork shape, as with Hubble's original diagram and in the image on this page. Elliptical galaxies form the handle of the tuning fork at the left. The sequence then separates to illustrate the two types of spiral galaxies. Irregular galaxies may be added at the right.

Although, as discussed, this is not an evolutionary sequence but simply a useful taxonomy of shapes, some astronomers continue to refer confusingly to those galaxies towards the left of the diagram as 'early type'.

The numbering system runs from E0 to E7 for the degree of ellipticity. Beyond E7 it becomes difficult to differentiate between a flat elliptical galaxy and a featureless spiral galaxy. Hubble himself realized he needed a transition type between elliptical and spiral galaxies. These galaxies are known as 'lenticular' galaxies and given the designation S0. A lenticular galaxy has the shape of a spiral galaxy but without any clear or obvious spiral-arm structure.

NGC 3923 – E4/S0

The example of an elliptical galaxy that we show below is NGC 3923. This has been classified as both E4 and S0 in various places. This is because it shows some of the attributes of an elliptical galaxy, but may alternatively be a lenticular galaxy seen almost face on.

Early images of this galaxy showed a fairly featureless structure, but the object becomes a lot more interesting at high resolution. As we see below, there appears to be a set of concentric rings of increasing radius emanating from the centre of the galaxy. The rings are particularly noticeable in this stereoscopic view. A ring in the interstellar medium of a galaxy can have a number of causes, including a collision where a smaller galaxy has passed through a larger galaxy and created a 'splash'. Indeed, astronomers think that the most likely cause for a series of rings like this is a series of collisions with dwarf galaxies, acting like stones falling into a pond and creating ripples. It might only be one or two dwarf galaxies that have each fallen through, oscillating back and forth through the middle of the galaxy, creating the rings.

NGC 3923 is classified as intermediate between an elliptical galaxy and a lenticular galaxy.

NGC 3115 seen from the Alston Observatory of the University of Lancashire, UK.

Furthermore, two extended features are coming out of the galaxy on the lower left of the image. The centre of the galaxy appears bright, and in fact the brightest part of the centre could be called elliptical - or it could be argued that it is more linear, or rectangular, in shape. It is a combination of these uncertainties that has led to the dual classification of this galaxy.

NGC 3115/Caldwell 53 – S0

NGC 3115 is seen almost exactly edge on. The image (previous page), taken at the Alston Observatory in Lancashire, shows the disc and bulge in perfect silhouette. This galaxy lies at a distance of about 30 million light-years, in the direction of the constellation Sextans. It is often classified as a lenticular galaxy, or S0.

The classification given to any object (such as lenticular, elliptical or spiral) depends on the relative brightness of the disc and the halo, the size of both, and the orientation of the galaxy. NGC 3115, being perfectly edge on, is a classic example of a lenticular galaxy.

Lenticular galaxies typically have a more massive halo than is found with spiral galaxies, but still have a disc. The most extreme case looks like a disc embedded in an elliptical galaxy. The least extreme case (which astronomers generally call S0) looks like an edge-on spiral, as is the case with NGC 3115.

What has happened with an S0 galaxy is that it has stopped forming stars in the disc. As a result, the disc fades rapidly in brightness because the most massive stars (which are the brightest) die out, leaving only the lower-mass stars. Consequently, the relative brightness of the bulge and the disc changes. Any remaining spiral structure fades in the disc, because such structure is more pronounced with the young stars than the older stars.

Lenticular galaxies in general, and S0 galaxies in particular, have always been thought of as a type of cross between spirals and ellipticals. We now go on to look at different types of spiral galaxies.

Normal Spiral Galaxies

The tuning-fork diagram shows two types of spiral galaxies. Those that have a linear feature in the centre are referred to as 'barred' spiral galaxies, and those that do not are referred to a 'normal' spiral galaxies. Barred spiral galaxies are designated by the letters SB, while normal spiral galaxies are simply designated by the letter S. Some astronomers designate normal spiral galaxies as SA, to distinguish them from SB, but we follow the original S. Note that the B (for 'bar') is a capital letter, in order to avoid confusion with the further sub-designations that we now describe.

Hubble decided to subdivide normal spiral galaxies into three categories, which he labelled a, b and c (lower case). The category depended on how tightly wound the spiral arms appeared to be, and on how prominent they were seen to be. So, a tightly wound spiral galaxy is designated Sa, a less tightly wound spiral is Sb, and a fairly open spiral is Sc. These latter Sc spiral galaxies are sometimes referred to as 'grand design' spiral galaxies, and they constitute some of the most beautiful objects in the night sky.

M64/NGC 4826 – Sab/Sb

The first example of a normal spiral galaxy that we show here is M64, or NGC 4826, which we see close to face on. This galaxy has been variously designated as intermediate between Sa and Sb (Sab) or as Sb. The spiral arms are fairly tightly wound, but because of the angle at which we are observing the galaxy, a dark lane appears to lie partly in front of the centre of the galaxy. Early images of this galaxy appeared fairly featureless until high-contrast techniques were used for display. At that point the dust lane became apparent.

The galaxy M64/NGC 4826 is classified as intermediate between an Sa and an Sb galaxy.

Our image shows not only the dust lane discussed, but also the detailed structure in the dust lanes. Furthermore, the spiral arms on the other side of the galactic centre are clearly seen. However, our image shows that the spiral arms are quite tightly wound, so we would agree with the classification of Sab or borderline Sb.

M77/NGC 1068 – Sb

Our next example of a normal spiral galaxy is M77, or NGC 1068. Its moderately open spiral-arm structure clearly places it in the category Sb.

The central region between the spiral arms is very bright, whereas the outer parts of this galaxy are fairly faint. So while most galaxies show a fairly uniform structure of brightness at the centre, becoming gradually fainter in a smooth manner, this galaxy does not.

The galaxy M77/NGC 1068 is classified as an Sb galaxy.

Instead, there is a jump in brightness, referred to as a 'discontinuity'. This galaxy is described in the *Hubble Atlas of Galaxies* (see Chapter 1) as the type example, or archetype, of an Sb galaxy with a brightness discontinuity.

The spiral arms show up partly as dark lanes on the face of the galaxy, but also in the form of a string of red circles. These are the places where new massive stars are being formed.

M77 is also what is known as an active galaxy, due to it having small radio jets at its centre. It is part of the class of active galaxies that are known as Seyfert galaxies (see Chapter 5).

M106/NGC 4258 – Sb

M106, or NGC 4258, is a spiral galaxy which lies at a distance of about 25 million light-years in the direction of the constellation Canes Venatici. It is classified as Sb and has clearly defined spiral arms, but it has irregular dust lanes, giving it a broken-up appearance. The nearside spiral arm is matched by one on the far side, but the arms are not symmetrical.

The spiral arms are quite straight and open, leading to this galaxy sometimes being classified as Sbc, or even Sc. The manner in which the spiral arms seem to end some distance from the bright galactic centre is almost indicative of a barred spiral galaxy. However, there is no evidence for a bar, so the origin of the structure is unclear. Likewise, there is no consensus as to why it is not symmetrical.

It is classified as an active galaxy because it has a black hole in its centre that is stirring up its surroundings. So this is another example of a Seyfert galaxy.

M106/NGC 4258, a spiral galaxy in Canes Venatici, about 25 million light-years away.

NGC 7331/Caldwell 30 – Sb

The galaxy NGC 7331 is classified as Sb. It lies at a distance of about 40 million light-years in the direction of the constellation Pegasus. It has been suggested as a galaxy that might resemble the Milky Way as seen from a distance. However, it has no bar, unlike the Milky Way, so we would question this comparison.

In the picture below, the galaxy appears to be slightly non-symmetrical, with one side extending further from the centre than the other. The galaxy centre appears bright yellow, whereas the spiral arms seem to be blue, which indicates new-born high-mass stars.

It is inclined at about 15 degrees away from exactly edge on. This means that we can see its spiral structure quite clearly and identify its type.

The galaxy NGC 7331, classified as an Sb galaxy, as seen by the Alston Observatory of the University of Lancashire, UK.

NGC 300/Caldwell 70 – Sc

We have chosen NGC 300 as an example of an Sc galaxy. It lies at a distance of about 6 million light-years in the direction of the constellation Sculptor. Its open spiral structure can be clearly seen. It has a relatively small central region with multiple spiral arms emanating from the centre. Each spiral arm separates into more than one spur. Indeed, the spiral structure is so open that it is sometimes classified as an Sd galaxy - that is, even more open than a typical Sc.

Dotted along the spiral arms we can again see red circles, which indicate regions of current massive star formation. There are also regions of blue that are visible along the spiral arms. These are areas where massive stars have already formed. We know this because the most massive stars are the brightest and bluest of the stars in any galaxy.

The galaxy NGC 300 is classified as Sc. The spiral pattern is so open that astronomers sometimes refer to it as an Sd galaxy.

M101/NGC 5457 – Sc

Another example of an Sc galaxy is M101, or NGC 5457, which is sometimes referred to as the Pinwheel galaxy. Again, it has a clear open spiral structure with multiple spiral arms coming from a bright centre, and each spiral arm separates into more than one spur. M101 lies at a distance of around 20 million light-years in the direction of the constellation Ursa Major, or the Great Bear. We saw an image of this galaxy in Chapter 2 taken through a radio telescope at a wavelength of 21 centimetres, showing low-density gas that is not energetic.

The first of the pictures (next page) was taken by the Hubble Space Telescope in visible light and shows a yellow-white bright centre, with the spiral arms appearing to be blue. The second image was taken from the ground by the Alston Observatory of the University of Lancashire. It was taken through a particular type of filter that transmits a narrow band of red light, known as a hydrogen-alpha filter. The filter shows energetic gas between the stars, which is the usual sign that new stars are forming there.

M101

The galaxy M101 is classified as Sc, shown here in an image from the Hubble Space Telescope.

The galaxy M101, as seen from the ground, through a hydrogen-alpha filter. This picks up the energetic gas where new stars are forming or are about to form.

A notable difference between the image above and the 21-cm radio image in Chapter 3 is that this image shows only a smaller bright centre, but with the spiral arms delineated by lines of red dots. These show where energetic gas lies. The radio image shows most of the rest of the gas in the galaxy, which is not forming stars, but occupies the space between the stars.

Barred Spiral Galaxies

Barred spiral galaxies comprise the lower branch of the tuning-fork diagram. Like normal spiral galaxies, they are classified by how tightly wound their spiral structures are. Thus tightly wound barred spiral galaxies are designated SBa, less tightly wound examples are SBb, and the least tightly wound are SBc. Astronomers believe that something like two-thirds of all spiral galaxies have bars.

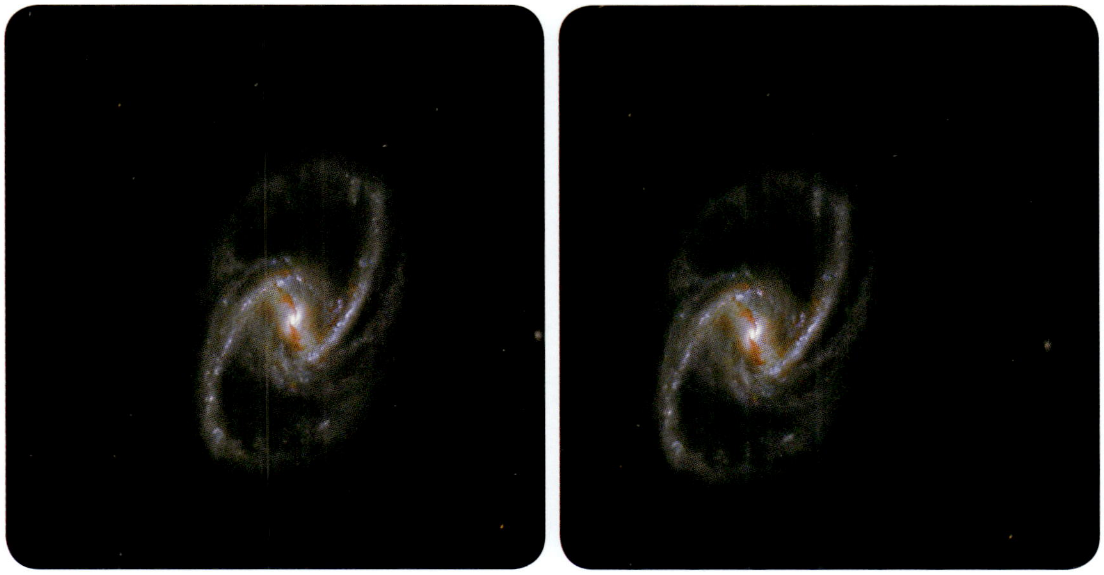

The barred spiral galaxy NGC 1365. It is classified as an SBb galaxy, and the bar can be seen clearly in this image.

NGC 1365 – SBb

Our first example of a barred spiral galaxy is NGC 1365. This is categorized as an SBb barred spiral galaxy. The linear feature stretching either side of the galactic centre is the bar. Two spiral arms radiate from the ends of the bar. The arms are not particularly tightly wound, giving this galaxy its designation as SBb.

The stereoscopic image (previous page) of the galaxy beautifully illustrates how the spiral arms unwind from either end of the bar.

NGC 613 – SBbc

NGC 613 is a barred spiral galaxy classified as SBbc, with the 'bc' indicating that the spiral arms are less tightly wound than those of NGC 1365. It lies at a distance of about 70 million light-years in the constellation Sculptor.

The centre of the galaxy is very bright and relatively compact. A straight bar appears from either side of the centre, and the two main spiral arms unwind from there. On one side there is an additional spiral arm that emerges from the end of the bar and then splits into two. On the other side a faint spiral arm emerges from the end of the bar and continues faintly around the outside of the galaxy. It is not straightforward to hypothesize how such a structure might have formed.

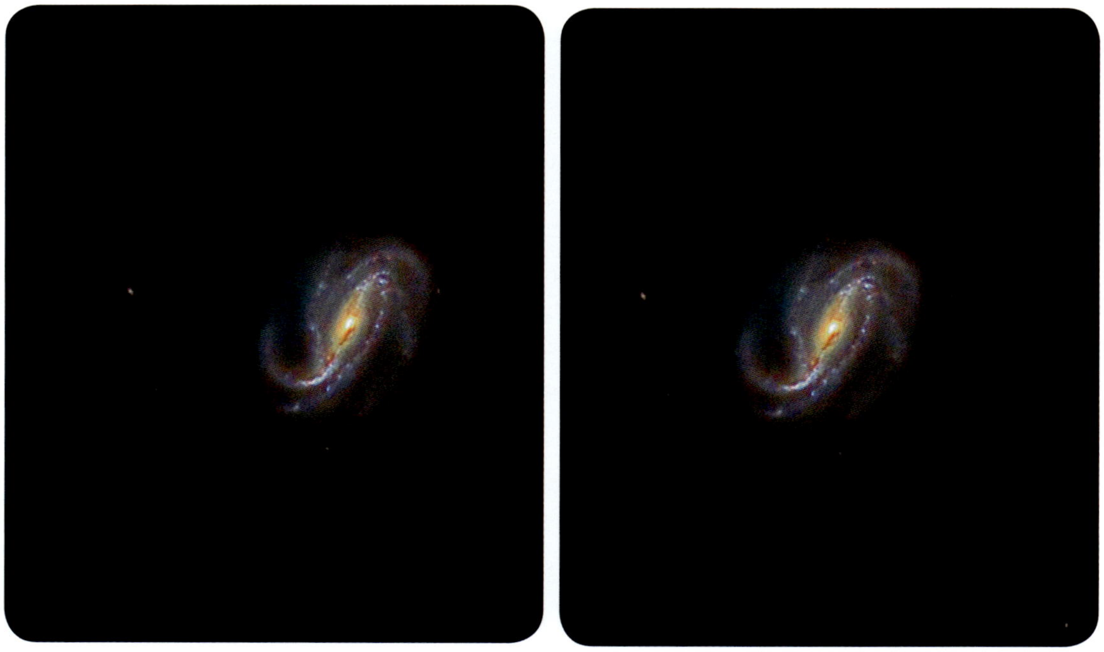

The galaxy NGC 613. It is classified as an SBbc galaxy, meaning that it is less tightly wound than NGC 1365.

M83/NGC 5236 – SBc

Finally in this section we show the barred spiral galaxy M83, or NGC 5236. This galaxy is designated as an SBc barred spiral. The central region contains a clear linear feature, or bar. Two spiral arms emerge from the bar in a wide fashion, which is not at all tightly wound. This galaxy shows a very large number of bright red circles, indicating where stars are currently forming. The red arcs consequently give this galaxy an overall reddish hue.

The galaxy M83. It is classified as an SBc galaxy, being less tightly wound than NGC 1365.

Irregular Galaxies

A galaxy can be classified as irregular for a number of reasons. The most obvious of these is simply that its shape does not fit into any other category. Consequently, the group of irregular galaxies can appear to be something of a ragbag of shapes, colours and sizes. Thus there is no such thing as a typical irregular galaxy.

The *Hubble Atlas of Galaxies* defines two types of irregular galaxies, which it calls Type I and Type II. In both cases the galaxies show no symmetry or obvious rotation axes. However, Irr I galaxies can be resolved into individual stars or clusters, whereas Irr II galaxies cannot, and instead appear rather smooth and featureless. Later astronomers questioned the usefulness of this separation, because, as telescopes became bigger and cameras and other instruments, along with their associated software, became more sophisticated, galaxies that had previously appeared featureless started to show features.

M82/NGC 3034 – Irr II

M82, or NGC 3034, is listed in the *Hubble Catalogue of Galaxies* as an 'archetypal irregular' galaxy of type Irr II.

M82 presents a rather splendid sight (next page). It almost appears to be two galaxies, but astronomers believe it is just one that has more than one axis. The main, long axis is of a bluish-white hue. We can see irregular markings against this in darker colours of orange and red. This axis could represent an elongated elliptical-type shape, but if so, it is more cigar-shaped than elliptical. Alternatively, it could be a more unusual shape, but seen edge on. In either case, it does not fit into any classification system.

Then there is a second 'axis' around which we see streamers of red and orange. This is now explained as a so-called 'starburst' galaxy, which is where a huge bout of star-formation activity takes place in the centre of the galaxy and blows apart a chunk of its surroundings, sending streamers of hot material thousands of light-years into space. A starburst galaxy may be triggered by another galaxy passing close by but not colliding with it. In the case of M82 the trigger is believed to be the nearby galaxy M81. We show a stereoscopic image of the two galaxies together (next page).

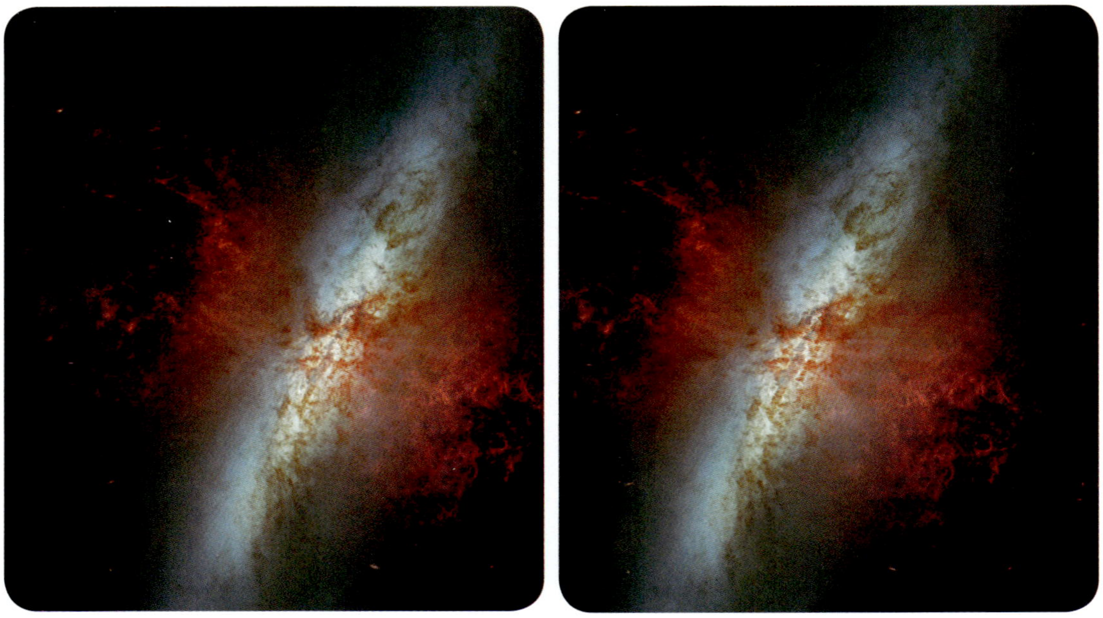

The starburst galaxy M82 in stereo (top, rotated) and mono (immediately above).

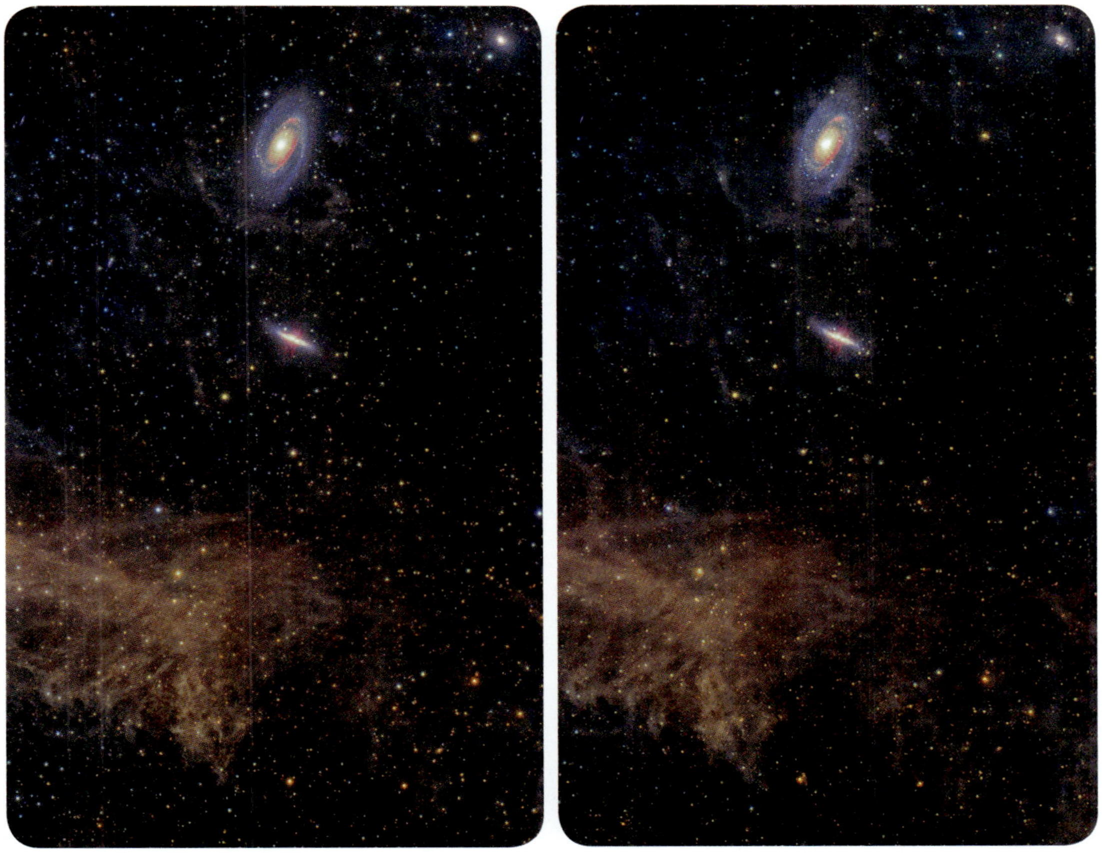

The galaxies M81 (at the top) and M82 (in the centre) seen together stereoscopically. It is thought that the interaction with M81 is what triggered the starburst in M82. The image shows them at a different orientation from the pictures opposite and below.

The galaxy M81, or NGC 3031 – see discussion on next page.

Grand Design Galaxies

The so-called grand design galaxies are those where the overall shape and structure are clear and symmetrical. So, grand design spiral galaxies are generally seen more face on than edge on, they have well-defined symmetrical spiral structures and generally look like how a galaxy 'should' look when you picture it in your mind's eye.

M81/NGC 3031 – Sb

M81, or NGC 3031, pictured on the previous page, is an archetypal grand design galaxy. It is classified as a normal spiral of type Sb. The central region is very bright and seen here as a yellowish hue. There are also dark lanes visible against the bright region. Further out the galaxy appears more purplish-blue, and very strong, clear spiral arms appear - as expected for an Sb galaxy - with a great deal of symmetry between the spiral arms, giving it the grand design status.

M51/NGC 5194, the famous Whirlpool galaxy, and its companion galaxy, NGC 5195. Some astronomers claim that the presence of a companion galaxy can generate a regular pattern in its neighbour, creating a grand design spiral pattern.

M51/NGC 5194 – Sbc

M51, or NGC 5194, is a grand design spiral galaxy. It is classified as Sbc and has clearly defined spiral arms, as in M81, but it has much more clearly defined dust lanes. Furthermore, there are a very large number of red regions dotted along the spiral arms like beads on a string that delineate the bright spiral arms just as clearly. These are regions of active and massive star formation and are HII regions (see Chapter 3).

Also seen in this picture is the companion galaxy NGC 5195, which is classified as an irregular galaxy. This galaxy is passing M51 and interacting with it, in that it seems to have pulled one of the spiral arms of M51 partly away from its host galaxy, and it now appears to follow NGC 5195. Some theories predict that a companion galaxy can help create and sustain a grand design pattern, rather like the way shepherd moons help define the rings and gaps in the rings around Saturn. Thus, NGC 5195 may be contributing to the orderliness of the grand design of M51. We return to this galaxy in Chapter 6.

M31/NGC 224 – the great Andromeda galaxy

The great galaxy in the constellation Andromeda, M31 or NGC 224, deserves a whole section of its own. It is the most distant object that can be seen by the unaided human eye (from a dark enough site and with sharp enough eyesight). It is the closest spiral galaxy to the Milky Way, at a distance of around two-and-a-half million light-years. It is a classic normal spiral galaxy of type Sb.

The Andromeda galaxy is currently heading towards us. The Andromeda and Milky Way galaxies are expected to collide – but in about five billion years, so nothing to panic about! They might pass through each other. Alternatively, they might orbit around one another and gradually merge into one giant galaxy, as we shall see with other galaxies later. Astronomers can model these collisions in simulations in supercomputers. However, there are too many unknown factors and variables to predict with any degree of certainty what form of interaction will take place between these two galaxies.

M31 is one of the most beautiful galaxies to observe. Its clearly winding spiral arms and distinct symmetry give it a very pleasing appearance, making it almost the archetypal form of galaxy. Even though it is seen at a much more steeply inclined angle than the grand design spirals discussed above, its spiral structure is still perfectly clearly delineated. The angle is 77 degrees away from face on, or 13 degrees from edge on.

The central bulge is very bright, and two spiral arms appear to be emanating from either side of the bulge. The nearside dark lanes are clearly delineated against the brighter central features.

The Andromeda galaxy

M31/NGC 224, the great Andromeda galaxy, is our nearest large spiral galactic neighbour, 2.5 million light-years away. Naturally it lies in the direction of the eponymous constellation. Inset: rotated stereo view of Andromeda.

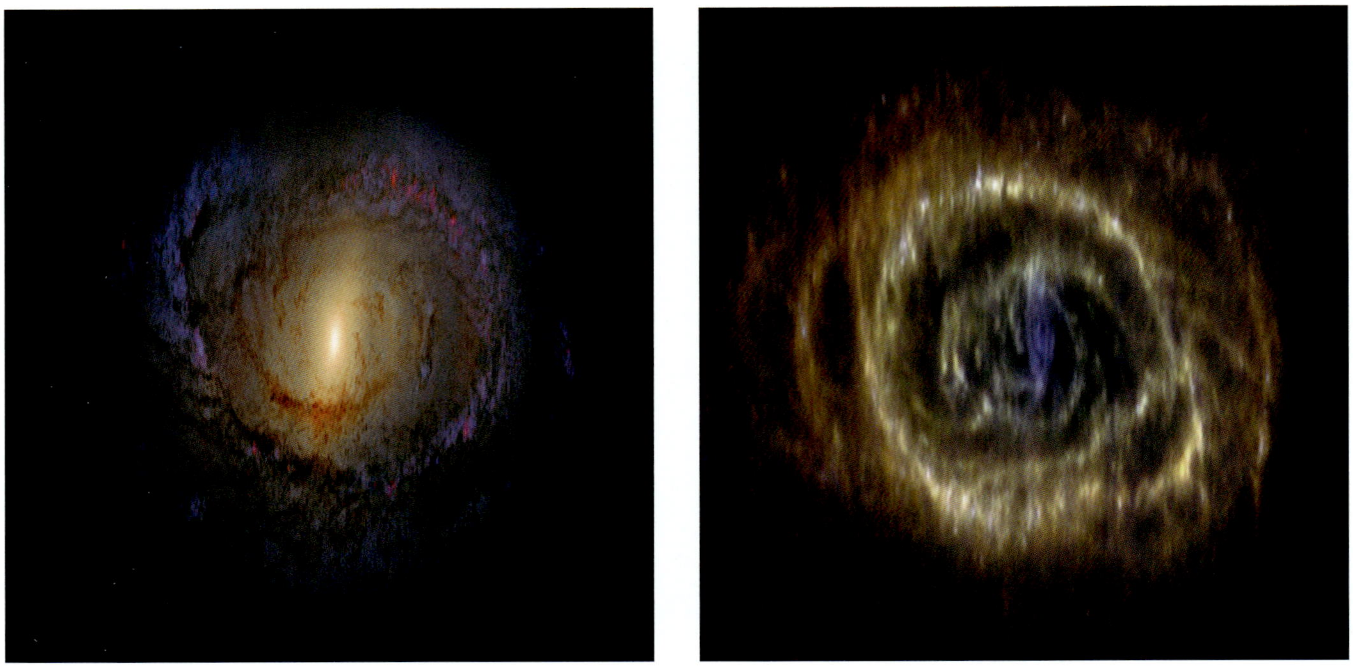

Left: the Andromeda galaxy, M31/NGC 224, de-rotated from its normal inclination angle of 77 degrees to appear face on in visible light. It is not clear whether it is a one-armed or a two-armed spiral. Right: the same de-rotated view but taken at far-infrared wavelengths by the Herschel Space Telescope, revealing what was concealed in the dark regions of the visible light picture.

Using specialist software, it is possible to predict what the galaxy would look like if de-rotated from its 77 degrees to appear face on instead. This gives a beautiful impression of the structure of the galaxy. However, it must be noted that certain assumptions go into this artificial de-rotation, so the image may not be completely accurate, although it is very suggestive.

The first thing to note is that it is less clear in this view whether the galaxy has a one-armed or a two-armed spiral structure. It is more or less possible to trace a single spiral from the edge of the bulge to the outermost edge of the galaxy. With this observation it is possible to go back to the original image and re-examine the spiral structure; it becomes possible to also see a single smooth spiral there too. Another thing to note in the de-rotated image is that the extension to the bright bulge region above and below is an artefact of the de-rotation process.

The de-rotated infrared image shows that perhaps the simple two-armed spiral is not as straightforward as it might appear. The far infrared is almost a negative of the optical picture, because dark lanes in the optical appear bright in the infrared. Now the pattern resembles a series of concentric rings in its outer parts. If this indicates the true structure, then the rings could be the aftermath of a smaller galaxy passing through the centre of M31 at some point in the past, sending out shock waves across the galaxy, rather like ripples on a pond. However, they may also be an artefact of the de-rotation process.

M31 is the largest nearby galaxy, but it is not the only one. A number of other nearby galaxies have also been identified, and together with M31 and the Milky Way, these are known as the Local Group of galaxies.

The Local Group of Galaxies

The Local Group is interesting to study because it contains a number of different types of galaxies, and all are close enough to allow us to study them in detail. The closest are the Magellanic Clouds, which are designated as satellite galaxies of the Milky Way. They were named in honour of the explorer Ferdinand Magellan but were known before then. Owing to their different sizes, they are known as (rather uninspiringly) the Large Magellanic Cloud (LMC) and the Small Magellanic Cloud (SMC). Both are classified as irregular galaxies, although because both have some semblance of a linear structure across their centres, it has been suggested that they be reclassified as barred galaxies.

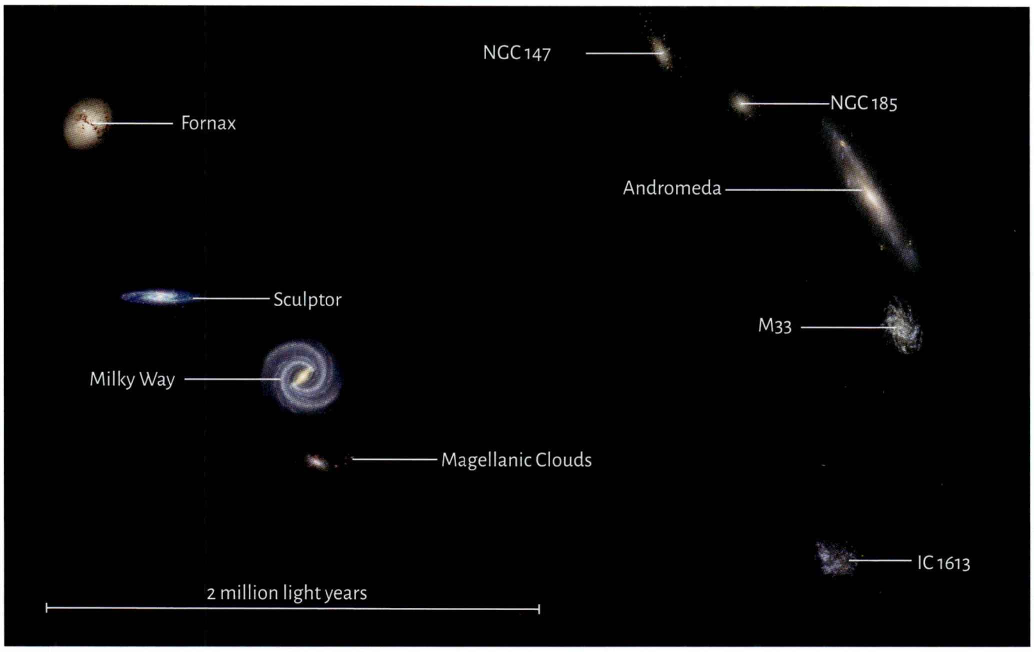

The Local Group of galaxies, showing the relationship between the Milky Way, the Andromeda galaxy and other nearby galaxies.

The Magellanic Clouds are dwarf galaxies that are gravitationally bound to the Milky Way. Here they are seen above the ESO VLT telescopes in Paranal, Chile.

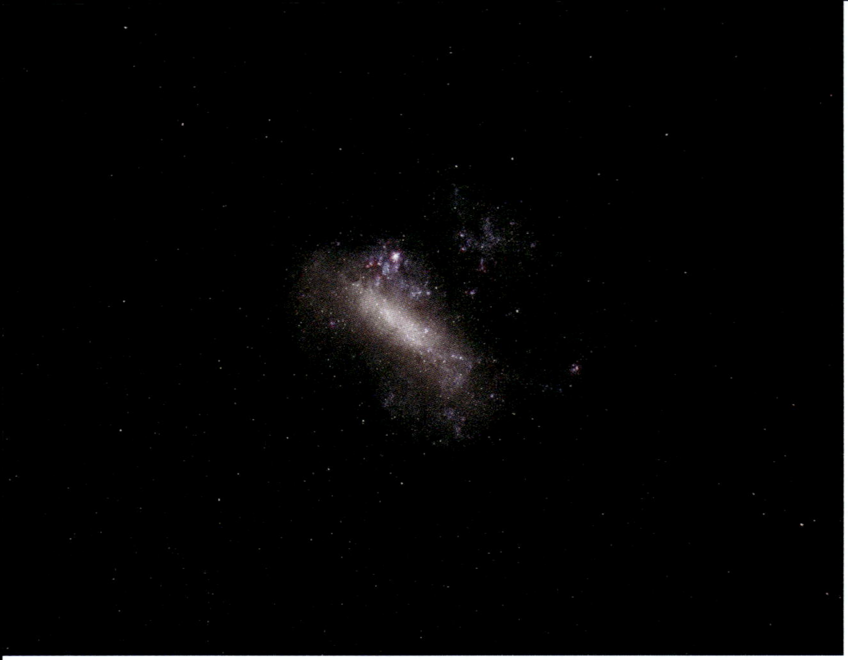

The Large Magellanic Cloud (LMC), taken from ESO La Silla Observatory in Chile using a DSLR camera. The LMC is a satellite galaxy of the Milky Way, visible from the southern hemisphere.

The Small Magellanic Cloud (SMC) as seen in the infrared by the VISTA telescope, allowing us to see through the foreground-obscuring dust. On the right of the image, the bright globular cluster 47 Tucana stands out.

The Sculptor dwarf galaxy lies at a distance of 13 million light-years and is known as a starburst galaxy for its current high rate of star formation.

The Fornax Dwarf galaxy contains many very old stars, as does the Milky Way. This supports the theory that the Milky Way was formed from smaller galaxies.

Other galaxies in the Local Group are the dwarf spheroidal galaxies in Sculptor and Fornax. Both are classified as E0, but neither looks very impressive in an optical image. Nevertheless, both have proved very useful to astronomers because of their proximity to the Milky Way. For instance, it is possible to examine the ratios of the chemical elements in greater detail within these galaxies than in those much further away. This type of study is known as the study of the 'metallicity' of the galaxy, and these galaxies can then be compared to the local values seen in different parts of the Milky Way.

Two other dwarf spheroidal galaxies in the Local Group are NGC 185 of type E3 and NGC 147 of type E5. They lie at about 2 and 2.5 million light-years away respectively. The pictures show how their ellipticities differ, and so how the classification changes in the range from E0 to E7 as galaxies become less spherical and more elliptical. However, it is important to remember that galaxies do not evolve to become more or less elliptical; this is simply a function of the physical parameters that they had when they were formed and of any mergers or interactions with other galaxies they may have undergone. We will explore this theme further in Chapter 6.

Metallicity

The fraction of chemical elements heavier than hydrogen and helium in a galaxy or region of a galaxy is known as its 'metallicity'. Straight after the Big Bang there were essentially no elements other than hydrogen and helium. Everything else has been made in the centres of stars by the process known as 'fusion'. Measuring the fraction of heavier elements can help determine how near to the pristine conditions of the very early Universe an object is, and how much its matter has been reprocessed in stars.

Top: NGC 147/Caldwell 17, a dwarf spheroidal galaxy of type E5 in our Local Group of galaxies.
Bottom: NGC 185, or Caldwell 18, a dwarf spheroidal galaxy of type E3 in our Local Group of galaxies.

IC 1613 is a dwarf irregular galaxy (image next page) just over 2 million light-years away that is also classified as a member of the Local Group of galaxies. The difference in shape between IC 1613 and the two elliptical galaxies is clear to see, showing why it is deemed to be an irregular galaxy. Slightly separate from the main part of the galaxy is a region of bright red emission. This marks the area where stars are currently forming. The red is marking out some giant HII regions (see Chapter 3), which are the markers of recent high-mass star formation.

One of the interesting differences between elliptical galaxies and spiral galaxies is that, while spiral galaxies have a great deal of material in their interstellar medium, elliptical galaxies generally do not. This means that while spiral galaxies are continually forming new stars, elliptical galaxies are not. The stars contained in elliptical galaxies are simply passing through their evolutionary stages over periods of billions of years. Some irregular galaxies are actively forming stars, such as IC 1613 as we have just seen, but some are not.

IC 1613 is a dwarf irregular galaxy in our Local Group. It is the bright yellow object just above the centre of the image.

One of the exceptions to this observation is M110, or NGC 205, which we saw above in the image of M31, the Andromeda galaxy. M110 is classified as a dwarf elliptical galaxy of type E5 (compare its apparent ellipticity with NGC 147 above). However, it is also classified as a peculiar galaxy (E5pec), because it has gas and dust in its interstellar medium. This is unusual for elliptical galaxies, as it means that this galaxy can continue to form stars. The proximity of M110 to M31 means that at some future stage it might fall into M31. If it were to do so, it would probably pass right through. In this case, it would almost certainly be stripped of its interstellar medium in the process, as we will see in other galaxies in the next chapter.

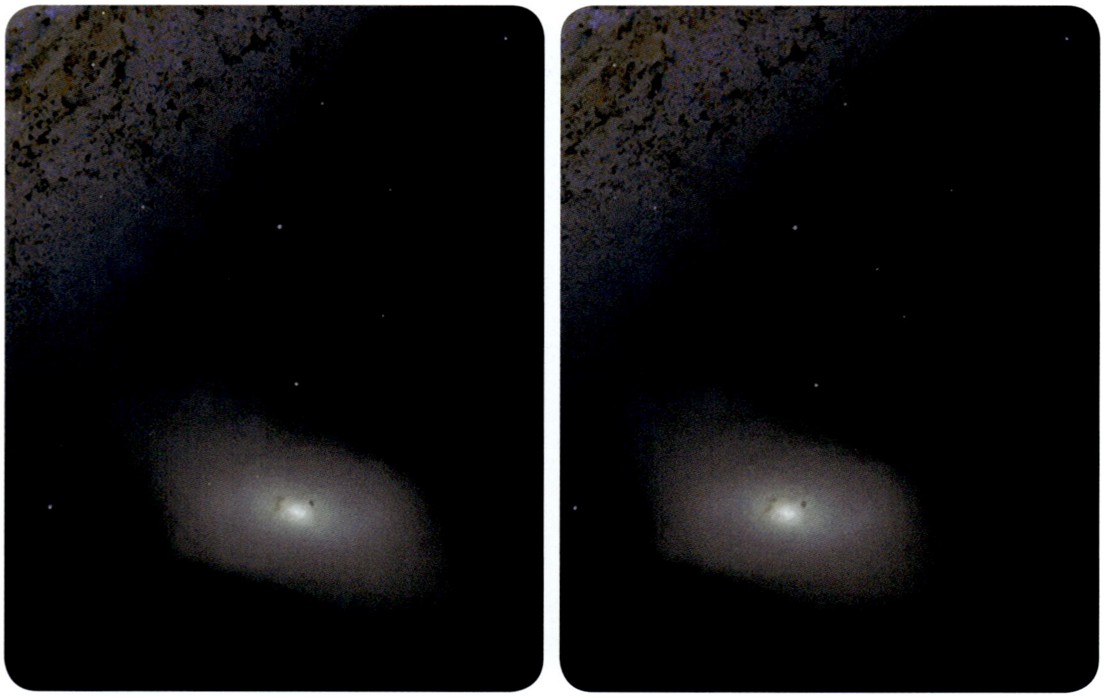

The dwarf elliptical galaxy M110 in our Local Group of galaxies. Its proximity to the Andromeda galaxy, M31, can be seen at the top left of the image.

We have already discussed the dominant members of the Local Group, Andromeda and the Milky Way, but there are two other members we want to mention. Messier 32, or NGC 221, is another satellite galaxy of the Andromeda galaxy – it can be seen above and to the left of Andromeda's centre in the large image earlier in this chapter. M32 is classified as an E2 galaxy, which is an elliptical galaxy that is fairly close to being spherical. It is also classified as a compact elliptical galaxy, due to its small size, earning it the additional classification cE2. It appears to have no interstellar medium, in common with most elliptical galaxies.

M32 is a dwarf elliptical galaxy in our Local Group. Its proximity to the much larger Andromeda galaxy, M31, can be seen.

The final member of the Local Group that we want to mention is Messier 33, or NGC 598. It is classified as an Scd spiral galaxy, because of its open and well-defined spiral arms. It lies in the constellation Triangulum and is the third largest member of the Local Group. It is slightly further away than the Andromeda galaxy, and a few people with exceptional eyesight can see it with the naked eye at a dark enough site. For these people, it just tops the Andromeda galaxy as the furthest object visible to the naked eye.

The central region of the galaxy is bright but fairly compact. It is a multi-armed spiral, and its spiral arms can be clearly picked out by the red circles strung out along the arms. These red regions are HII regions (see Chapter 3) and sites of recent star formation. They give the overall galaxy a slightly pinkish glow against the backdrop of the already formed bright blue stars. Study of the velocity structure of this galaxy and its rotation curve (see Chapter 2) implies that it may have a very large dark-matter halo for a galaxy of its size.

M33/NGC 598, the Triangulum galaxy, is a spiral galaxy that is part of our Local Group of galaxies, the second-closest galaxy to our own.

However, observations of the motions at the centre of the galaxy by the Doppler shift method (see Chapters 2 and 5) suggest that, if it does indeed have a black hole at its centre, then it must be at least a thousand times less massive than the black hole in the Milky Way Galaxy (see Chapter 3). We can thus conclude that the size of the black hole at the centre of a galaxy does not scale with the size of the dark matter halo around a galaxy. That presents challenges for theoretical models of the manner in which galaxies form, such as the model we discussed in Chapter 2.

The Virgo Cluster

We have dubbed galaxies as 'islands in infinity', but to misquote John Donne, no island is alone. It came as a surprise to astronomers when it was discovered that the whole of our Local Group of galaxies was moving at a rate of a few hundred kilometres per second in the direction of the constellation Virgo. Specifically, it is moving towards a group of galaxies known as the Virgo Cluster. Clusters of galaxies are defined as groups of galaxies that all come within the same sphere of influence in terms of gravity. They are also known as a 'self-gravitating' collection of galaxies.

Thus astronomers came to the conclusion that the Milky Way and the Local Group must represent an outer part of the Virgo Cluster. So we are genuinely not alone in that sense, and we see that these islands in infinity are all, in fact, part of island chains, each forming its own archipelago. We will return to the topic of clusters of galaxies in Chapter 7.

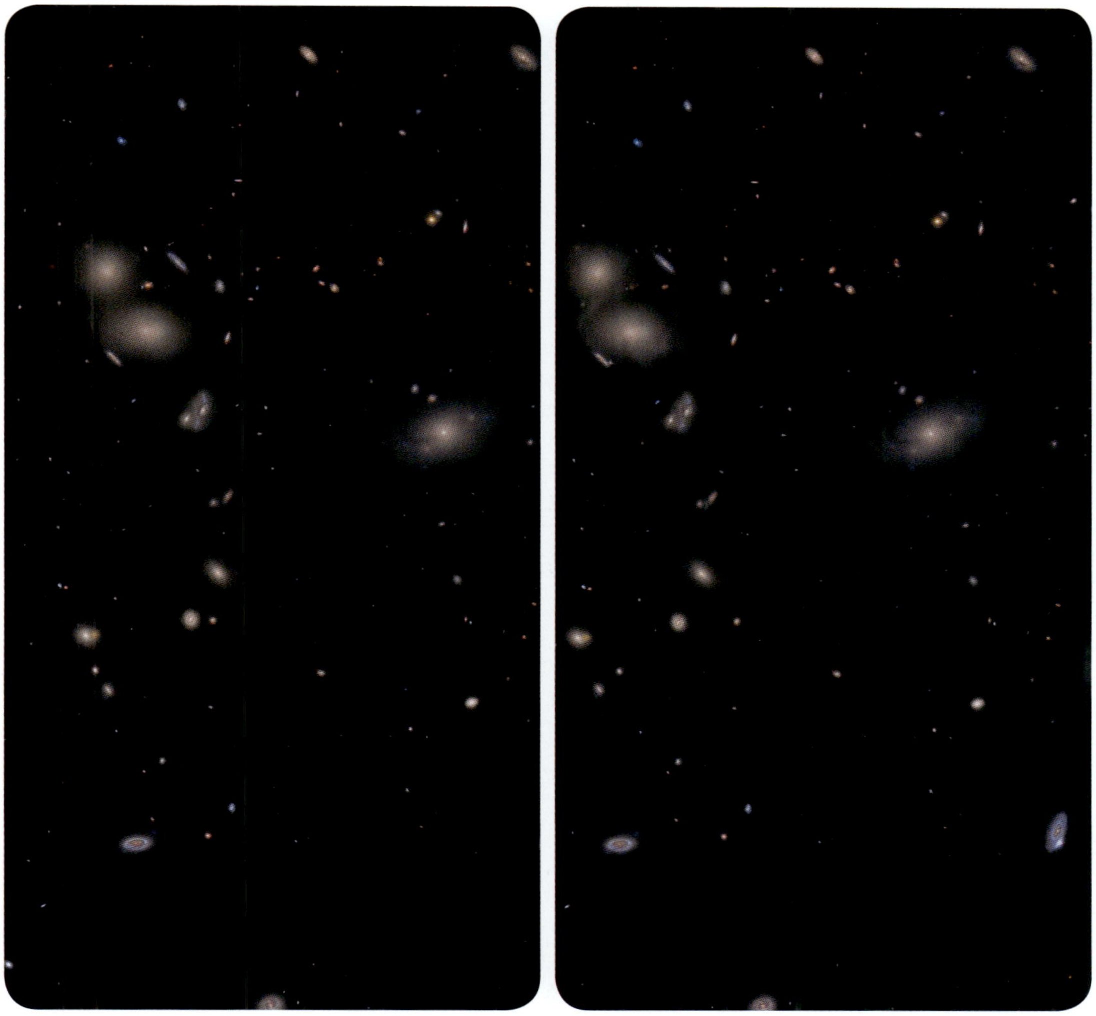

Our Local Group of galaxies is actually an outer part of the Virgo Cluster of galaxies, whose inner part is seen here.

This is the centre of the Virgo Cluster of galaxies. It lies at the heart of the bigger Virgo Supercluster. Our Milky Way, together with the Local Group, is an outlying member of the Virgo Supercluster.

The Flying Saucer galaxy

NGC 4565 is an edge-on spiral galaxy. Both the disc and bulge can be seen.

Galactic Discs and Bulges

5

Components of Galaxies

Galaxies are the principal building blocks of the visible Universe, but they are composed of smaller constituents. The most obvious of these are the stars that shine and give a galaxy its brightness and colour. The next most obvious components are the dark lanes we see in all pictures of galaxies. These are evidence of the material between the stars, which astronomers call the interstellar medium (ISM). As discussed in Chapter 3, the ISM is made up of gas and dust. The gas is mostly hydrogen and helium, with traces of heavier elements. The dust is what causes the appearance of dark lanes, and consists mostly of silicates and carbonaceous material (carbonates and hydrocarbons).

When we think of silicates we tend to think of sand, rather like the sand on a beach, but the silicates in the ISM are much smaller particles. The particles of ISM dust are approximately the size of particles of smoke, and just as smoke blocks out light - which is obvious if you see light rays from a window passing through smoke (see next page) - the dust in the ISM blocks out the starlight behind it. Smoke can also scatter light, making it sometimes appear bright against a dark background, as we see in the photograph, and we also see this effect in some of the galaxy images.

Other components of a galaxy that are important are the radiation field, the dark matter and the magnetic field. The radiation field comes from the combined emissions of all the stars in a galaxy. The dark matter was discussed in Chapter 2. The magnetic field comes from a combination of sources, including the magnetic fields of all the stars and planets, and from the charged gas particles of the ISM, which are perpetually moving around (when a charge moves, it creates an associated magnetic field).

Why Interstellar Dust Affects Our View of Galaxies so Strongly

The dust grains in the ISM are roughly the same size as smoke particles on Earth – the majority ranging from 1/10 of a micrometre up to 1 micrometre across, although a few much smaller and much larger particles are present too. The visible light that we use to view these regions has wavelengths from about 0.4 (blue) to 0.8 (blue) micrometres (microns), which is very much in the same range as the dust particles. When light interacts with aggregates of particles, the strongest scattering effects occur in exactly this situation, where the wavelength of the light matches the size of the particles. This is why the interstellar medium often presents us with such spectacular images.

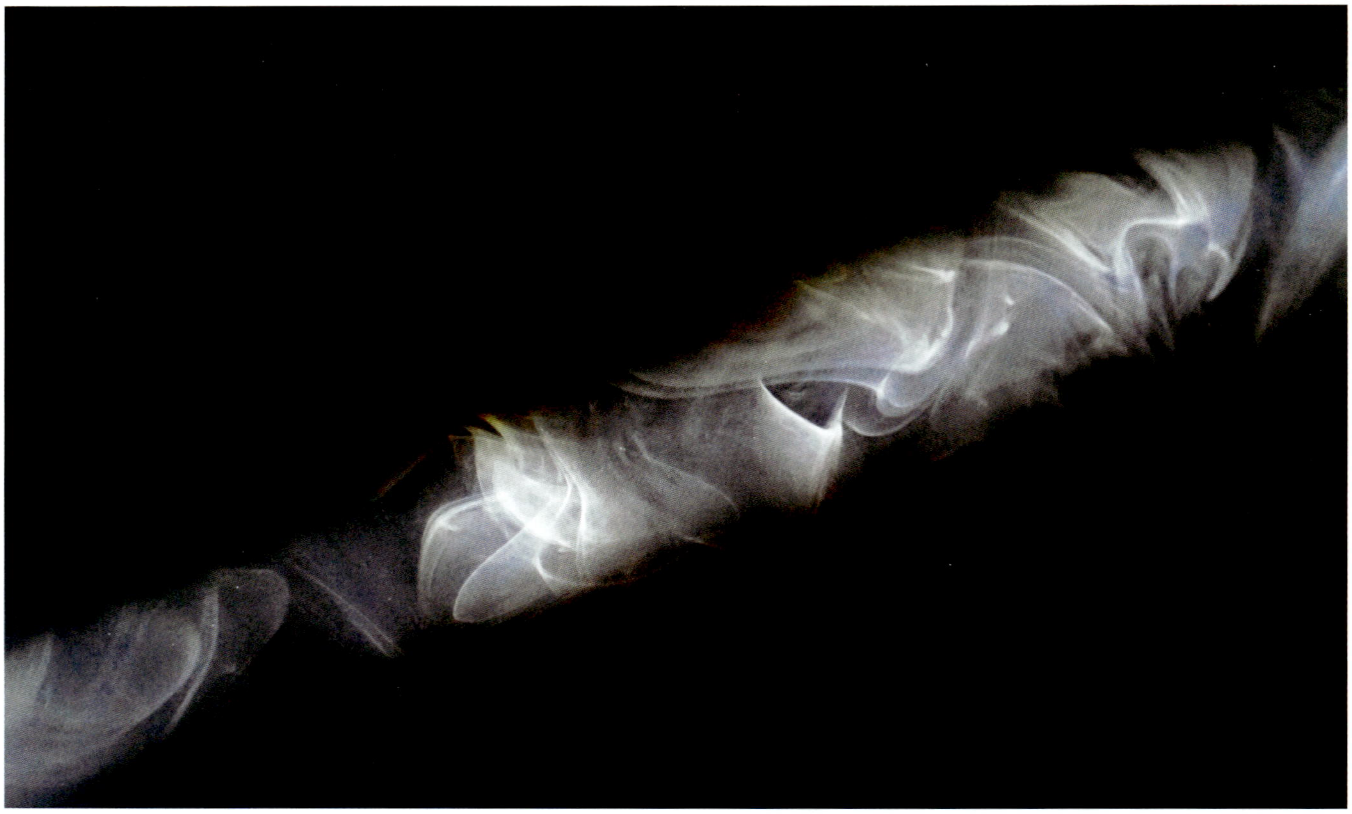

To illustrate our point regarding dust, here we see terrestrial smoke illuminated by light from a window. The smoke will obscure anything that is behind it, but in this situation, the swirling eddies in the smoke particles are caught in the shaft of light, and, being comparable in size with the wavelength of the light, they scatter the light – in all directions, but most significantly towards us, forming the bright rings and filaments that we observe here. These patterns are very reminiscent of the features we see in our photographs of dust clouds in galaxies.

Galactic Structures

The various components we have just discussed are spread throughout a galaxy in different ways. For a disc galaxy, such as a spiral or a barred spiral galaxy, the schematic diagram opposite shows how the structure of a galaxy is made up. The largest component is of course the disc.

Astronomers tend to subdivide the disc into a thin disc within the central plane of the galaxy and a thick disc surrounding it. The thin disc is where stars are actively forming today, and the thick disc contains slightly more evolved stars. The relative sizes and importance of the two vary from galaxy to galaxy. A good example of the thick and thin discs can be seen later in this chapter in NGC 891. We discussed the bulge in our Milky Way Galaxy in Chapter 3.

Surrounding all this is a region known as the galactic halo. This is a large, extended region where globular clusters are located. Globular clusters are believed to form some of the first stars in a galaxy and hence contain the oldest stars. On an even larger scale, surrounding everything is the dark matter discussed in Chapter 2.

At the very centre of a galaxy there is probably a black hole. Surrounding this, on a scale of possibly a thousand light-years, is the part of the galaxy that astronomers refer to as the 'bulge'. The relative size and importance of the bulge varies from galaxy to galaxy. Studying the bulge tells us about the galaxy's formation and past evolution.

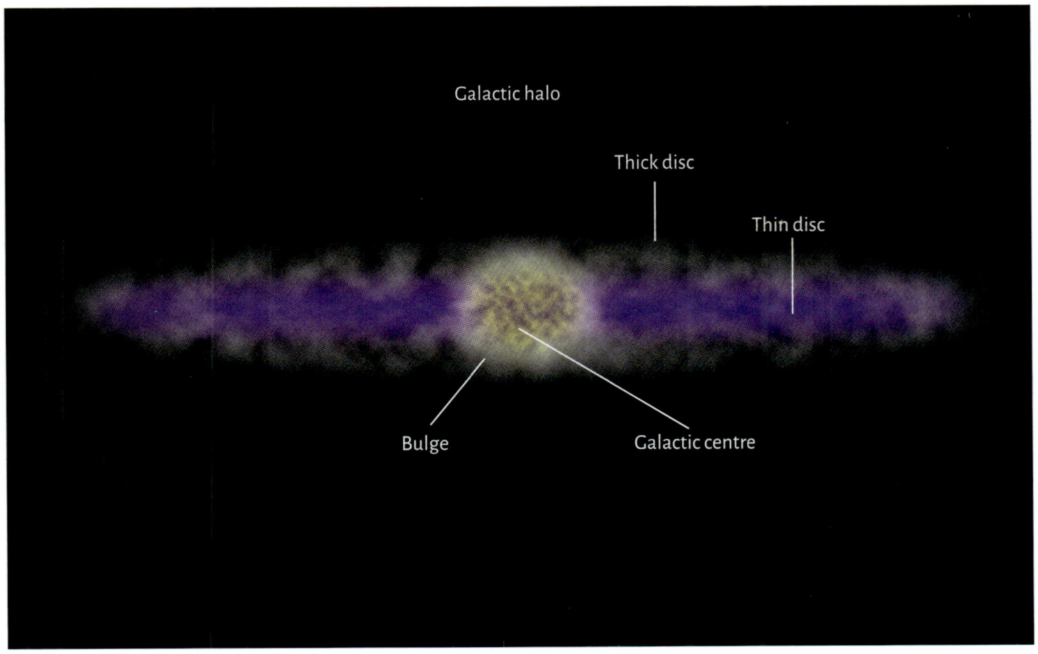

A schematic of an unbarred spiral galaxy seen edge on.

Studying Edge-On Galaxies

The best way to study the bulges of galaxies is to investigate galaxies that appear to us edge on. In this way we see the bulge from the side, and can tell how large and pronounced it is in each case. This, after all, is the angle at which we see our own Milky Way Galaxy, which makes edge-on galaxies the easiest to compare to the Milky Way. We previously described the different shapes of bulges, including 'boxes' and 'peanuts', both of which are non-spherical, the difference being that peanuts have a more pinched 'waist'.

We will now look at images of some edge-on galaxies and try to assess the sizes of their bulges, and hence the relative significance of their bulges compared to their discs. We start with the least pronounced bulges and progress through ever larger bulge sizes, remembering again that this is not an evolutionary time sequence, but merely a convenient classification system. Those galaxies with the least pronounced bulges are referred to as 'disc-dominated' and those with the most pronounced bulges as 'bulge-dominated'. We start by considering those with little or no bulge.

Disc-Dominated Galaxies

NGC 55/Caldwell 72

Our first example is NGC 55. This was originally classified as an Sc galaxy (see Chapter 4 for galaxy classification definitions), meaning that it is believed to have a well-developed spiral structure. Astronomers can deduce this from the structure of the dark lanes in the disc, seen against the brighter background regions. Some astronomers have argued that there is evidence for a bar in the galaxy and that it is actually an SBc galaxy. Indeed, *The Cambridge Photographic Atlas of Galaxies* classifies it as SB. The diameter of the disc is roughly five times the thickness of the thick disc.

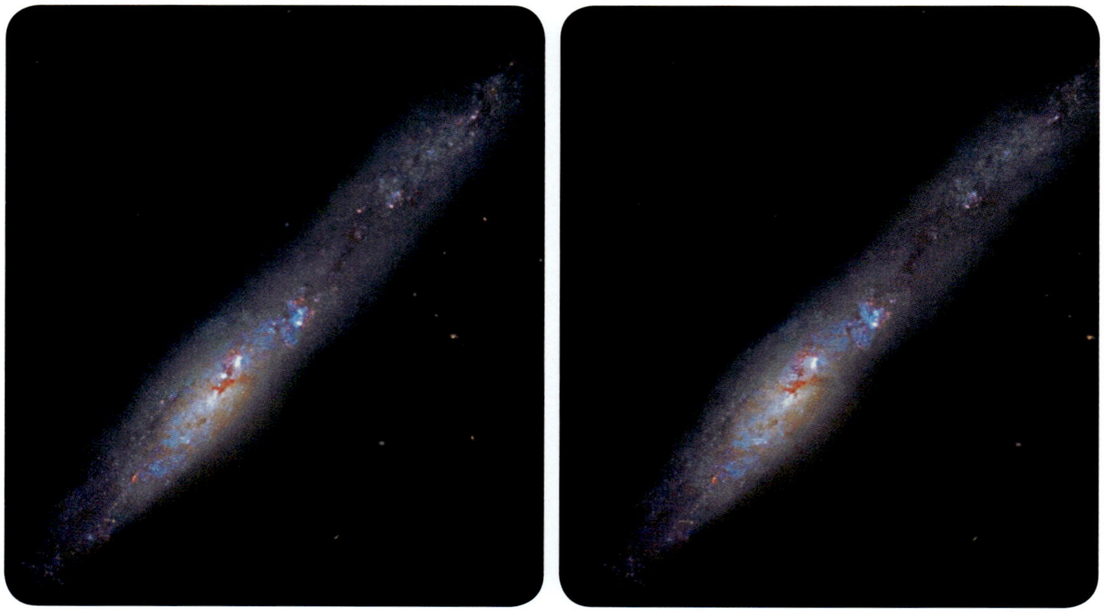

The galaxy NGC 55. Note how the brightest part seems offset from the centre.

One unusual aspect of this galaxy is that the brightest region of the galaxy, believed to be the galactic centre, appears to be offset from the centre of the disc. Some astronomers have argued that this gives cause to suggest that, if we could see this galaxy face on, we would classify it as irregular because of its lack of symmetry.

Furthermore, there is no obvious bulge visible in this galaxy. Thus we can say that any bulge the galaxy might have can be no wider than the thickness of the thick disc, and that if there is a bulge around the galactic centre, it is offset from the centre of the disc, perhaps because of a previous interaction with another galaxy.

These observations about the galaxy give us some clues about how it formed and evolved. The lack of a central bulge may indicate that the galaxy formed from a more diffuse or dispersed initial cloud of gas and dust, so that no central condensation was able to form. Alternatively, it may have formed from a cloud that was rotating quite rapidly – astronomers would say it had large angular momentum. The rapid rotation could have flattened out any central bulge.

Another possibility is that the galaxy underwent a collision with another galaxy, which stripped away any original galactic bulge that may have been there. A collision might also explain why the galactic centre is off centre with respect to the disc.

Study of the galaxy in stereo allows us to estimate which components of the galaxy are in front of which other components, and which are behind, giving us a three-dimensional picture of the large-scale structure of the galaxy.

Further work involving measuring the Doppler shift (see box on next page) of different parts of the galaxy would enable us to tell which parts are moving towards us and which are moving away, allowing us to make a dynamical model of the galaxy. Putting all of this information together might enable astronomers to determine which of the proposed scenarios about the formation and evolution of this galaxy is correct.

NGC 253/Caldwell 65

NGC 253 is classified as an Sc galaxy, with quite an open spiral structure. It is orientated about 10–15 degrees from edge on. It is further classified as a special subset of the class, or Sc(s), because the spiral pattern is most clearly seen in its dark lanes, rather than in bright spiral arms. Our viewpoint allows this to be seen against the background bright starlight, and also in the disc of the galaxy. Only around the outside edge do we see a clear, bright spiral arm.

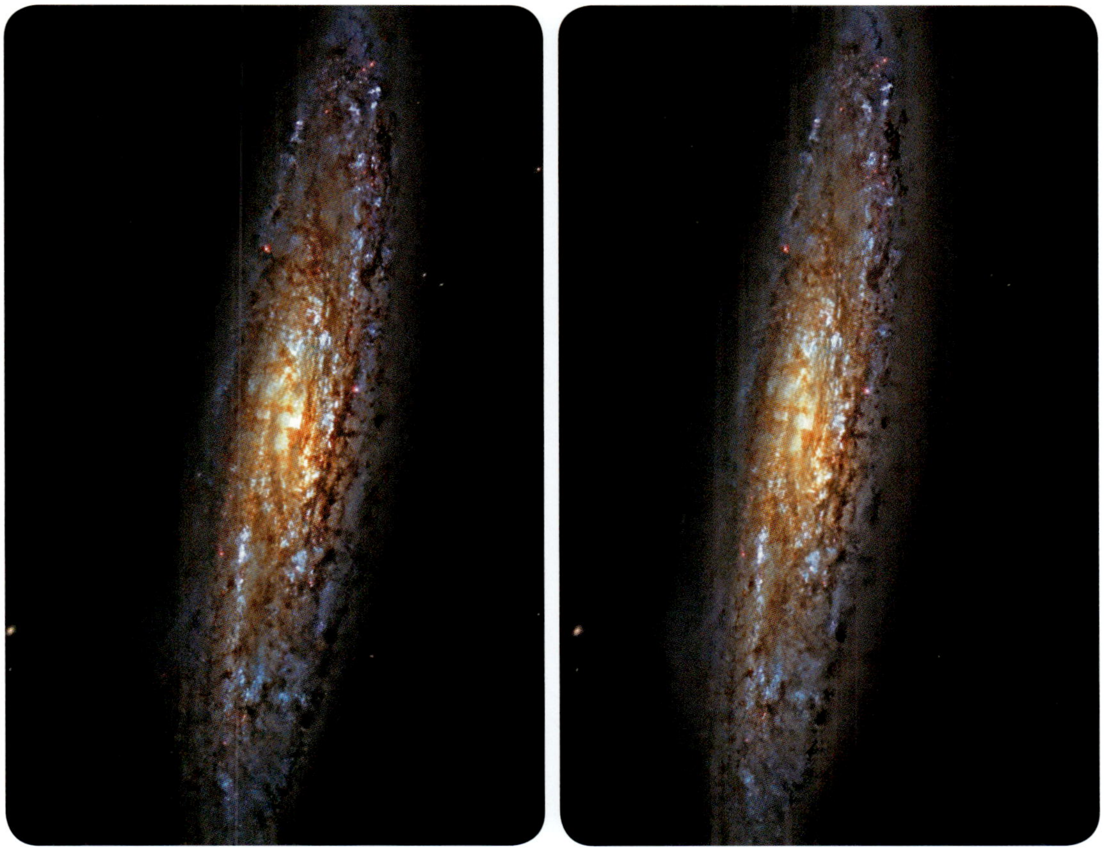

The galaxy NGC 253. The spiral dark lanes can be seen clearly in front of the centre of the disc.

However, once again we see no evidence for a strong, distinct bulge, although the centre of the galaxy is very bright compared to the rest of the galaxy. So, as with NGC 55, we are left to speculate whether this galaxy formed from a more dispersed cloud of matter, with no central condensation.

It is only when we look at NGC 253 stereoscopically that we really see the spiral pattern standing out strongly. This example shows the power of stereoscopy in the analysis of astronomical images particularly well. Suddenly the dark lanes appear clearly in relief. Furthermore, the thick disc is really apparent in the stereoscopic images, and if anything, the centre of the galaxy looks narrower than the thick disc.

The Doppler Shift and the Motions of Stars

To us humans, the stars appear to be fixed in the sky. If we observe the closest of them for a number of years, we might detect some 'proper motions', but even then the movements are very small. The stars in distant galaxies do not appear to move, even over centuries, but we can infer their motions by measuring shifts in the colour of the light we receive from them (see the diagram in Chapter 2).

We can record the spectrum of a star or group of stars, and look for a shift in the wavelengths of features in the spectrum. If we find that the lines in the spectral patterns have moved towards longer wavelengths (toward the red) we can infer that the stars are moving away from us. In simple terms, this is because the waves have got stretched on their way to us. From the magnitude of this 'redshift' we can estimate the speed of the motion. Similarly, if the spectrum is shifted to shorter wavelengths (blue), we can measure the speed at which the object is travelling towards us. Such shifts are known as Doppler shifts. If we compare the Doppler shifts of a number of groups of stars in a distant galaxy, we can tell how each group is moving relative to the rest, and so build up a picture of the dynamics of the galaxy.

NGC 4945/Caldwell 83

NGC 4945 is also classified as an SBc galaxy, as are NGC 55 and NGC 253, although with a bar. Once again we see very strong dark lanes that stand out particularly clearly in the stereoscopic images. However, there is now very slight evidence for a bulge, both above and below the plane of the galaxy. There is a distinct glow on either side of the plane. This glow is widest at the centre of the galaxy. Particularly above the galactic centre, there is a hint of a bright streamer of material. The galactic centre itself is very bright compared with the rest of the galaxy.

The galaxy NGC 4945. Again, the dark lanes can be seen in front of the galaxy disc.

However, the single most striking thing about this galaxy is that it is not symmetric about its centre. On one side the disc is very bright and has a bright yellowish colour. On the other side the disc is much less bright and generally much more reddish. Bright yellow stars are generally much younger than redder, fainter stars.

What this probably means is that there has recently been a large outbreak of star formation on the side of the galaxy where we see large numbers of bright yellow young stars. This outbreak does not extend to the other side of the galaxy, where the stars are all older and redder.

We can only speculate about this lack of symmetry. Why one side of the galaxy has formed a large quantity of young new stars and not the other is not easily explained. One possibility is that another galaxy may have flown past NGC 4945 on one side, creating a gravitational disturbance as it went by. This could have triggered the formation of stars on that side of the galaxy.

NGC 4244/Caldwell 26

NGC 4244 appears as an elegant needle-like shape, but it is in fact a spiral galaxy seen edge on. It is about 15 million light-years away in the constellation Canes Venatici (the Hunting Dogs). It is classified as an Sc galaxy, although some have questioned this, given that no spiral structure can be seen clearly in this completely edge-on galaxy.

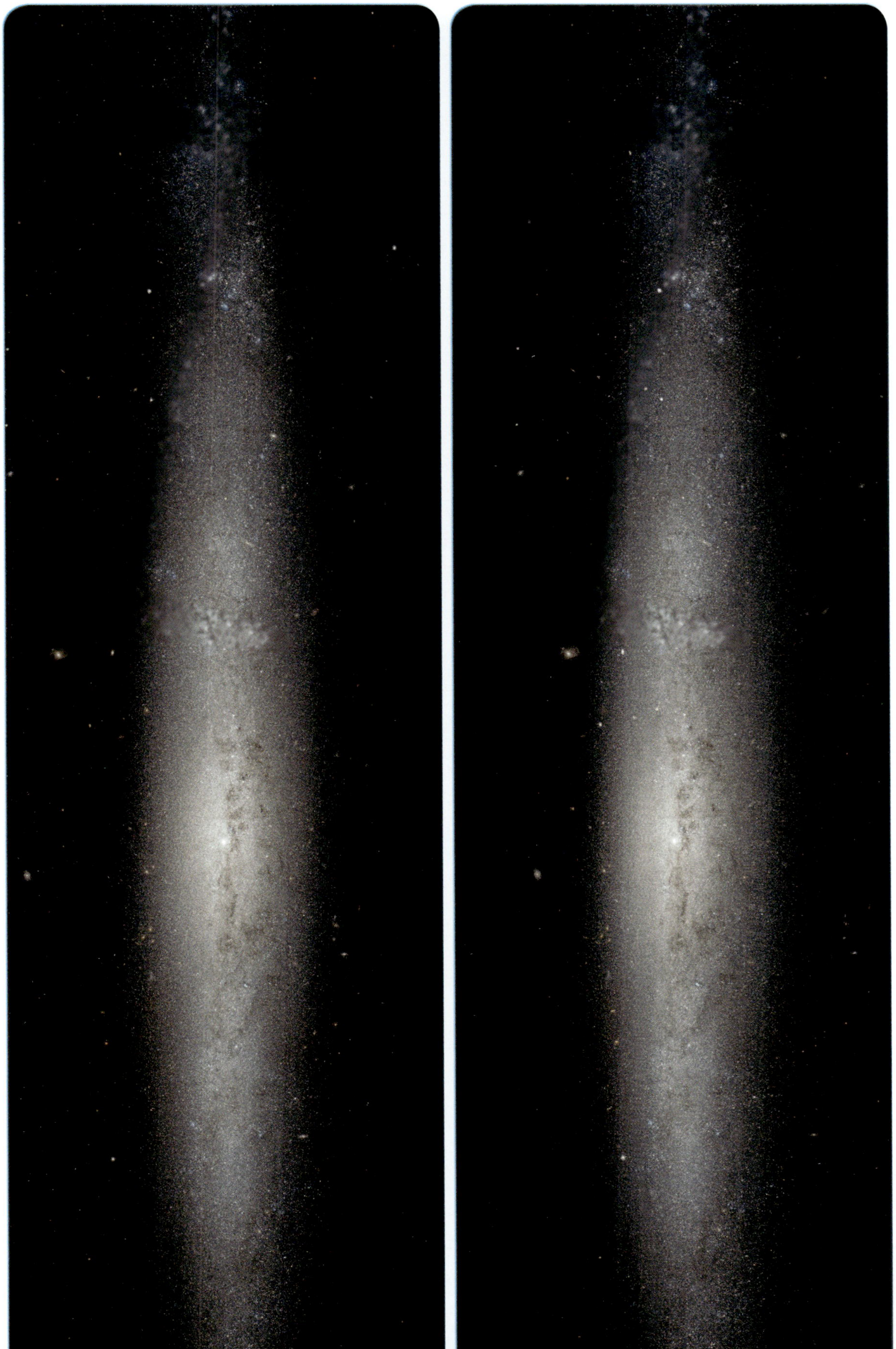

A Hubble Space Telescope image of the Silver Needle galaxy (NGC 4244) converted to 3-D.

There is a dust lane visible on one side of the centre of the galaxy, partially crossing the bright central point. This has been taken as evidence of a spiral arm. The dust lane appears to break up further from the centre, perhaps as the spiral arm disappears around the far side of the galaxy.

However, on the opposite side to the dust lane, the galaxy seems almost completely smooth. Astronomers are debating why this might be so.

The image above was taken by the Hubble Space Telescope, but in 2008 the Gemini Observatory in Hawaii discovered a massive star cluster in the centre of this galaxy that was rapidly rotating. A cluster such as this is known as a nuclear star cluster (NSC) due to its proximity to the centre, or nucleus of the galaxy. It has been observed that the more massive the galaxy, the more massive the NSC it can host.

Some evidence has also been found of a small, faint halo around the centre of this galaxy, although it is not apparent in the picture. Thus we have seen that this galaxy has an NSC and a small halo, but no evidence whatsoever of any galactic bulge.

NGC 3628

NGC 3628 is another edge-on spiral galaxy. It lies about 35 million light-years away in the constellation Leo. In fact, it is part of the Leo Triplet of galaxies, along with M65 and M66. It is sometimes known as the Hamburger galaxy due to its resemblance to a burger in a bun seen from the side.

It is classified as an Sb(pec) galaxy, meaning an unbarred spiral galaxy with a peculiar feature. Some astronomers have argued that it is a barred spiral galaxy, with the bar seen end on, but not all agree with this interpretation.

There is a very clear, broad dust lane right across the middle of the galaxy, which is cited as evidence of a spiral galaxy. The ends of the dust lane appear to curve away from the plane of the galaxy. This is the peculiar feature that gives the 'peculiar', or 'pec', label. Such curvature is known as warping of the disc. It is a relatively common feature of disc galaxies, but this is a particularly clear example.

The other clear features of the galaxy, apart from the dust lane, are the bright linear features above and below the dust lane (the bun around the burger). This is an elongated, flattened bulge.

As we mentioned when discussing the Milky Way Galaxy, some galactic bulges are described as boxy, peanut-shaped or even X-shaped. When observed at other wavelengths, the flattened bulge of NGC 3628 is X-shaped. This has led some astronomers to speculate that there is a bar in the centre of this galaxy, just like the one in the Milky Way.

NGC 3628

The galaxy NGC 3628 is an edge-on spiral galaxy. The dark lane of the galaxy is seen to curve slightly in the lower part of this image, which is known as a warp. This galaxy has what is known as a tidal tail (see Chapter 6), which can just be made out as a faint feature extending off the bottom of this picture.

Some theories of how bars form in disc galaxies involve interactions with nearby galaxies. NGC 3628 is part of the Leo Triplet, so interactions with the other two members, M65 and M66, could be responsible for the bar, if its presence is proved.

Similarly, interactions with nearby galaxies can cause the sort of warping we see in the disc of NGC 3628, and there is a dwarf galaxy in its tidal tail. So the appearance of NGC 3628 could be largely due to its environment.

Milky-Way-Like Bulges and Discs

NGC 891/Caldwell 23

NGC 891 is classified as an Sb galaxy, and it is seen completely edge on. Now there is definite evidence of a bulge at the centre of this galaxy. It clearly shows the classical disc galaxy shape of two saucers, one inverted on top of the other, placed rim to rim.

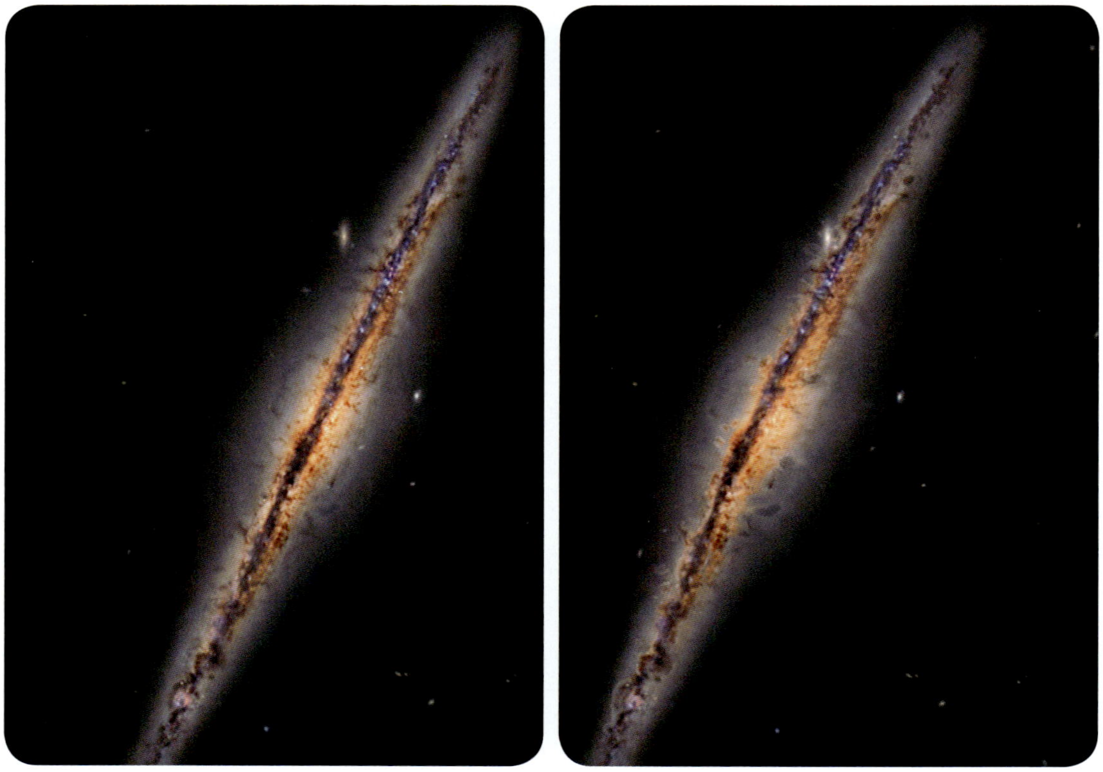

The galaxy NGC 891. This galaxy appears to be similar to the Milky Way.

The dark lane almost perfectly bisects the centre of the galaxy and is continuous from one side of the galaxy to the other. The thickness of the dark lane is a few hundred light-years. The bulge in this case is clearly visible, while still appearing relatively small. The thick disc of stars that we referred to at the start of this chapter is also clearly seen on either side of the thin disc, delineated by the dark lane.

The bulge in NGC 891 has been compared to the bulge in our own Milky Way Galaxy, which of course we also see edge on. The implication is that if we headed out of our Galaxy in a direction exactly away from the Galactic Centre, but still staying in the plane of the Milky Way, until we were well outside our Galaxy and looked back, we would see something very much like NGC 891. Unfortunately, we are currently unable to prove this one way or another, but it is an interesting thought.

NGC 3079

NGC 3079 is a galaxy lying at a distance of about 50 million light-years in the constellation Ursa Major, which we see almost, but not quite, edge on. It has been classified as SBc, an open-armed barred spiral galaxy, although this is somewhat uncertain due to the angle at which we see the galaxy.

The galaxy NGC 3079 as seen by the Hubble Space Telescope.

Structures resembling spiral arms can certainly be seen across the disc in the Hubble image. They are particularly clear on the near side of the disc, where at least two concentric spirals are visible.

At its centre a supermassive black hole has been discovered, which is roughly 2 million times the mass of the Sun, rather like the black hole at the centre of the Milky Way. Yet, unlike the Milky Way, NGC 3079 has very little, if anything, in the way of a bulge at its centre.

Close examination of the central region of the picture shows a kind of bubble apparently being blown out to one side from the centre of the galaxy. This is the usual sign of an active galactic nucleus and is a signature of the presence of a black hole at the centre of a galaxy.

When we zoom into the centre of the picture, we see a truly spectacular explosion, almost like an eruption from an active volcano. Matter appears to be being shot out of the galaxy to a distance of many hundreds or even thousands of light-years.

Seyfert Galaxies

There are several types of active galactic nuclei. Among the most common are Seyfert galaxies, first named by Carl Seyfert (1911–60) in 1943. They are characterized by a bright, compact, active region in the very centre of each galaxy, a black hole at the very centre, and highly energetic outflowing matter coming from the centre. They are essentially lower-mass versions of quasars.

The Chandra Space Telescope took an X-ray image of the same region at the centre of NGC 3079. This is equally spectacular and shows that the matter must be at extremely high energies because it radiates in the X-ray region – it is probably highly charged and moving at extremely high speeds in a very strong set of magnetic fields.

We now know that NGC 3079 is a Seyfert galaxy, and we see energetic material driven out in an explosive manner, as we would expect with such a galaxy.

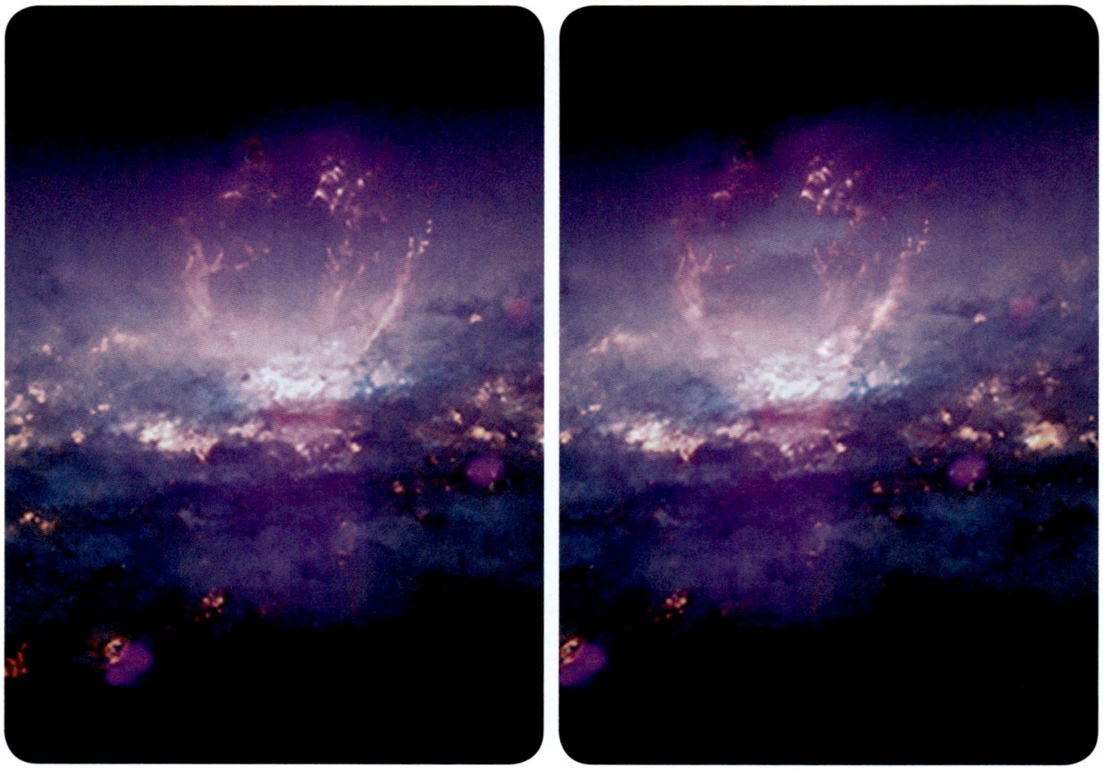

A close-up of the bubble at the centre of NGC 3079 seen in X-rays.
This is known as an active galactic nucleus (AGN).

NGC 1055

NGC 1055 lies near the celestial equator in the constellation Cetus, often known as the Whale (although the translation is actually 'sea monster'). It lies in a small group of galaxies that includes Messier 77 (which we discussed in Chapter 4). They are at a distance of roughly 50 million light-years.

NGC 1055, again seen very nearly edge on, is classified as an SBb galaxy. In other words, it is believed to be a barred spiral galaxy, with quite an open spiral structure. The edge of the disc can be seen darkly against the background of the rest of the galaxy.

Galaxy NGC 1055

Galaxy NGC 1055 seen edge on. This large barred spiral galaxy is nearly 116,000 light-years across, up to 15 per cent larger in diameter than the Milky Way. The bright ball of light in the centre is the galaxy's bulge. Dark lanes can be seen in front of the bulge. These are the spiral arms of the galaxy. In 3-D we can see the structure in the arms much more clearly.

In the image, the bright reddish-orange central region of the disc shows a hint of a bar-like structure. Furthermore, the outer bluish parts of the disc show a just-discernible open spiral structure. These factors combine to give the galaxy its SBb classification.

A variety of differently coloured regions can be seen across the disc. The brighter, bluish regions show where new stars are still forming or have recently formed. This is because young stars are generally brighter and bluer than older stars.

The bulge of the galaxy appears as a diffuse, milky-white structure around the central regions and is slightly more pronounced than that seen in NGC 891. It is almost as if the bulge is merging into the halo. Furthermore, the bulge appears somewhat asymmetrical - that is to say, on one side of the disc the bulge appears more spherical than on the other side, which appears flatter.

NGC 4565/Caldwell 38

NGC 4565 is classified as an SB galaxy and is also seen very close to edge on. If we performed the same experiment as we described for NGC 891 and left our Milky Way Galaxy in the same way and looked back, it is possible that our Galaxy would look more like this.

NGC 4565 is another galaxy that is similar to the Milky Way.

The bulge here is slightly larger relative to the rest of the galaxy. The dark lane is once again seen bisecting the galaxy, particularly against the galactic bulge. The bulge itself appears yellowish orange, especially when contrasted with the bluish hue of the outer parts of the disc.

Also note how the centre of the galaxy is relatively bright, and that the galaxy gets fainter as we look towards the outer edges. This is an effect with two main causes. Firstly, the centre of the galaxy is generally more densely populated, so the stars nearer the centre of the galaxy are closer together and there are more of them in a given volume. Secondly, stars generally form at a faster rate nearer to the centre of the galaxy, so there are commonly more young stars nearer the galactic centre, which are in turn brighter than their older counterparts.

Bulge-Dominated Galaxies

M104/NGC 4594 – the Sombrero galaxy

M104 is classified as an Sa/Sb galaxy, and is again roughly edge on, although very slightly more inclined than the previous example, at an estimated angle of about 6 degrees. The classification of edge-on galaxies is somewhat uncertain because we are not able to see their spiral structures clearly. In this case the galaxy is given a dual classification to recognize the uncertainty.

The galaxy contains an extremely bright galactic centre, surrounded by a very large extended bulge, seen as a large bright area surrounding the galactic centre and extending roughly halfway to the edge of the disc.

Again, a very clear dark lane can be seen right around the edge of the disc. Some structures can be made out in the disc itself, showing as clumpy dark areas against the brighter background structure of the disc.

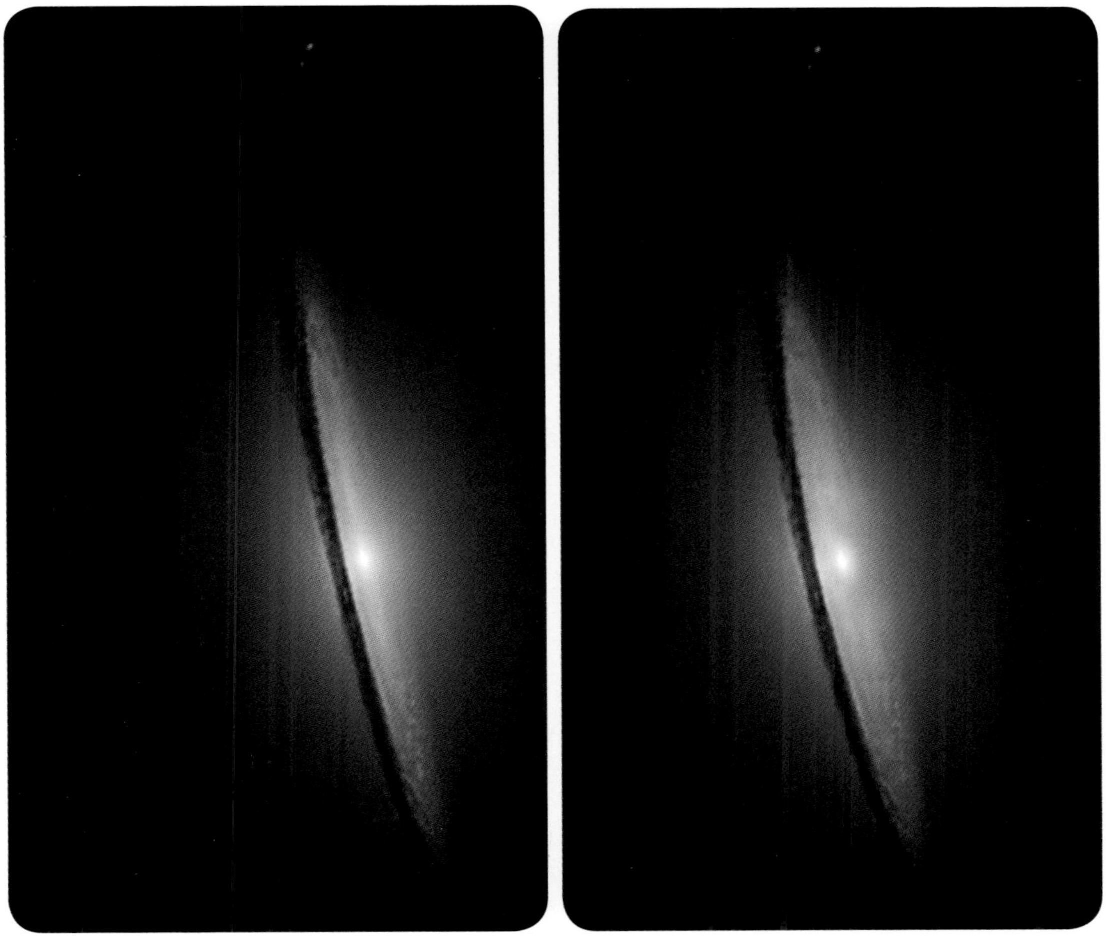

The galaxy M104. The bright bulge is clearly seen in the centre of the galaxy, but note also the extended faint glow stretching either side of the galaxy. This is the outer part of the bulge.

M104, the Sombrero galaxy, is one of the most beautiful edge-on galaxies visible in the night sky. The disc of the galaxy is clearly delineated by the dark lane, the bright centre of the galaxy is visible just above the dark lane, and the extended bulge can be seen either side of the disc.

The stereoscopic images very clearly show the dark lane encircling the outer ring of the disc. This is another clear example of how stereoscopy can reveal more information about a galaxy than a simple picture.

M104 is often referred to as the Sombrero galaxy because of its shape resembling the hat of that name (or even two hats, with one inverted and both placed rim to rim). It is a favourite of galaxy photographers due to its very large extended bulge.

The brightest region at the centre of the galaxy is relatively small when compared to the large extended structure of the bulge, which is much fainter. This indicates an uneven distribution of stars within the bulge of the galaxy, with a large number of bright stars near one another in the centre, surrounded by a much lower density of stars, almost creating two distinct regions.

We also show a false-colour image taken by three NASA space telescopes, Chandra, Hubble and Spitzer. Collectively, these were known as NASA's Great Observatories. They worked at completely different wavelengths, X-ray, optical and infrared. This allows us to get three very different views of the galaxy, which are shown on the right of the image (next page).

A false-colour image of the galaxy M104, where each colour represents a different wavelength: X-ray (blue), optical (green) and infrared (red). The right-hand column shows the three component images that are blended together to make up the main image.

The Chandra X-ray image (in blue) traces the hot gas in the galaxy. The Hubble optical image (in green) shows the light from stars. The Spitzer infrared image (in red) displays the extent of the cool dust in the interstellar medium.

Notice how the different components of the galaxy are brightest at different wavelengths. The X-rays only trace the central part of the galaxy, and the inner region of the bulge. This shows the hottest parts of the galaxy. The optical image shows the sombrero shape most clearly. It shows the stars in both the disc and the bulge.

Also note how the dark lane of the edge of the disc is only visible in the optical image. This reflects what we said at the start of this chapter about interstellar dust being most efficient at blocking optical light.

The infrared shows the disc at its brightest, since that is where the dust lies. So the regions that are dark in the optical glow brightly in the infrared. The infrared image also shows brightness in the central regions of the galaxy, where it is picking up not only the dust, but also the old red stars.

NGC 7814/Caldwell 43

The galaxy NGC 7814 can be found in the constellation Pegasus at a distance of just over 50 million light-years. It is somewhat similar in appearance to M104, in that it is seen almost completely edge on, has a bright centre, and has an extended bulge that seems to merge into its halo. As a result, it is sometimes referred to as the Little Sombrero. The bulge completely dominates the image of the galaxy, with a disc seen as a very thin dark line almost exactly bisecting the bulge.

The galaxy NGC 7814. The way in which the dark lane almost perfectly bisects the disc can be clearly seen in this stereo image.

NGC 7814 is classified as type Sab, meaning it has no bar, but is a classical spiral with fairly tightly wound spiral arms. Once again, our view is from a position close to the galaxy's principal plane, and the dark lanes in its disc can be seen clearly against the background of the disc and bulge. This disc has a vaguely orange glow, the centre appears white, and the bulge has an almost bluish tinge to it. The size of the visible disc is about 85,000 light-years in diameter. There is a school of thought among astronomers that suggests a very large bulge, such as this one, implies a relatively young galaxy. In this hypothesis the galaxy NGC 7814 is still in the process of forming, and the bulge is still collapsing onto the disc, making this as young as only about a billion years. However, not all astronomers agree with this idea.

Galaxy Evolution

One question that many astronomers want answered is whether we can deduce anything about the evolution of galaxies from studying their shapes. For instance, we mentioned above the hypothesis that a galaxy with a bulge almost as big as its disc might be in an earlier evolutionary state (younger) than one with a much smaller bulge.

The theory is that the bulge is collapsing onto the disc under the influence of the gravitational pull of the disc. It is an idea that dates back to Hubble himself. Unfortunately, it does not always stand up to closer scrutiny.

The way in which we can test this is by studying the motion of the gas and stars within the galaxy. We study the motion using the Doppler shift, as described earlier, to detect whether the matter is falling towards the centre of the galaxy or accelerating away from the centre, or doing something more complex.

IC 2051 is a spiral galaxy with a large bulge and bar of stars slicing through its centre. It was observed in visible and infrared wavelengths as part of a Hubble Space Telescope study on galactic bulges in 2019.

What astronomers have found in disc galaxies is generally that the overall rotation of the disc is the dominant motion, as we saw in Chapter 1. Studies of bulge-dominated galaxies have proved inconclusive about the nature of the evolution of disc galaxies. Studies of the internal motions of elliptical galaxies often show highly complex motions that are not easily explained by any one simple form of evolution.

This has led astronomers to hypothesize that perhaps galaxies start out as various forms of disc galaxies. The extremely deep images taken by Hubble and the JWST appear to show that disc galaxies, rather like those we see around us today, are present in the very early Universe. These galaxies then collide, and the resultant galaxies are either elliptical or irregular. This hypothesis would suggest that the most important driver of the evolution of galaxies is the process of galaxy collisions. This was thought to be a rare process until relatively recently but is now known to be more common than previously expected.

The Antennae galaxies

NGC 4038 and NGC 4039 are a pair of galaxies in mid-collision. The image is dotted with blue star-forming regions, surrounded by hot hydrogen gas, which appears in pink.

Mergers and Chaos

6

When Galaxies Collide

We have called galaxies 'islands in infinity', but as we saw in Chapter 1, they are actually relatively close together compared to their sizes. We have found that galaxies exist in groups and clusters, and that very few, if any, galaxies exist in pure isolation. They may be islands, but they exist mostly in island chains and archipelagos. Furthermore, nothing is static in space; everything is moving. Hence collisions between galaxies are fairly common, producing some of the most beautiful images in the cosmos.

In this section we look at what happens when collisions occur and in the aftermath of such events. In some cases the images are quite magnificent, but in others we see the cosmic equivalent of train wrecks. In all cases we learn more about the physical conditions within galaxies, because this is the only way we can, so to speak, dissect a galaxy and see its innards spread out across space for our inspection. We end by including oddities and things that astronomers struggle to explain.

Near Misses and Glancing Blows

When galaxies do not quite collide but pass close by to each other in a near miss, these are not strictly collisions. Astronomers label them as 'interactions' because the gravitational effects of each galaxy can have consequences for the other. Gravity is a long-range force that can act across light-years of space, meaning that galaxies do not have to directly collide to affect each other.

M51/NGC 5194 and NGC 5195

We previously saw M51/NGC 5194 in Chapter 4, where we described it as a grand design spiral galaxy, classified as Sbc, with clearly defined spiral arms. We mentioned that it has a companion galaxy, NGC 5195, which is an irregular galaxy, passing close by.

We must be careful when claiming that two objects which appear to be next to each other in our pictures are closely related, as they could be at very different distances. Apparent closeness that is not real in three-dimensional space is known as a 'projection effect'.

However, this is not the case here. We can see the spiral arm of M51 in front of the galaxy NGC 5195, and it looks as if NGC 5195 is pulling that spiral arm outwards and generally interacting with M51. We also mentioned that NGC 5195 may be contributing to the orderliness of the grand design of M51, in a similar manner to the shepherding moons of the rings of Saturn.

The galaxy M51, or NGC 5194, and its companion galaxy NGC 5195. Note how the spiral arm on the right of M51 is being pulled away by the companion galaxy.

Astronomers have used very large supercomputers to model what might have happened in this case. From the results of these tests, they have come up with a likely scenario for the intersection between these two galaxies, which is that the smaller galaxy (NGC 5195) has passed by M51 and is now moving away.

In the process, NGC 5195 has deformed one spiral arm of M51. This can be seen by looking at the end of the spiral arm on the opposite side of M51 from the smaller galaxy. Then draw an imaginary line from that point through the centre of M51 to the spiral arm nearest to NGC 5195. It can be seen that the spiral arm does not stop there but continues on to the location of the smaller galaxy, NGC 5195. This is due to gravity from the smaller galaxy pulling out this spiral arm, which is known as a 'tidal' effect. It is sometimes referred to as 'tidal stripping' if any matter is stripped from the larger galaxy. In this case it looks as if the extended spiral arm might well be being stripped off M51.

If we were to observe this galaxy again in, say, 100 million years' time, M51 might look quite different, with one spiral arm stripped away a lot more. At that point the smaller galaxy, NGC 5195, might well have moved on so far that it would be impossible to tell that it was this galaxy that disrupted M51.

NGC 3718 and NGC 3729

The galaxy NGC 3718 lies at a distance of just over 50 million light-years in the constellation Ursa Major, the Great Bear. It is part of a cluster of galaxies known as the Ursa Major Cluster (we discuss clusters in Chapter 7).

It has been variously classified as Sa(pec) or S0(pec) or irregular – where 'pec' stands for peculiar. We can all agree that it looks peculiar, but what it is and how it attained this shape have been subjects for debate.

The galaxy NGC 3718. Observe how the spiral arms have been pulled straight by an interaction.

The early classifications are flawed because all that astronomers could detect was the bright central region with a dark lane. Hence the S0 (borderline elliptical) or Sa (spiral without clearly defined, open-patterned spiral arms) original classifications. The dark lane extending across the central region was the feature that led to the 'peculiar' classification.

However, the *Hubble Atlas of Galaxies* discusses the 'faint extended features' as if they might be spiral arms and speculates about whether the dark lane could be a dark bar, even though bars are usually bright. It goes on to note that if the dark lane were a bar, the galaxy could potentially be classified as an SBb galaxy (an open, barred spiral).

Some astronomers have claimed that the dark lane is a ring around the centre of the galaxy (based on 21 centimetre radio wavelength observations) and that the galaxy should be classified not as a barred spiral but as a polar ring galaxy (see later in this chapter). Our stereoscopic image gives us much room for conjecture.

The galaxy NGC 3718 with companion NGC 3729 (on the right), which may have caused the spiral arms to be straightened by its gravitational attraction as it passed by.

Now that we can see the galaxy in all its glory, we realize that there are two straight 'spiral' arms, with one extending from each side of the bright, central region. Each one ends in a curly, wispy structure that seems to envelop the entire galaxy. In this image the SBb classification seems entirely plausible.

The unusual straightness of the spiral arms leads to the question of how they came to be like this. Some astronomers believe that the shape arises from a previous interaction with a passing galaxy, such as we saw happening to M51 (earlier in this chapter).

The galaxy believed most likely to have been involved in the interaction is NGC 3729, which lies only 150,000 light-years from NGC 3718. Astronomers think that NGC 3729 may have passed by NGC 3718 at some time in the past, disrupting what might have looked like an ordinary barred spiral galaxy. We do not know, but maybe in the future M51 will evolve to look something like NGC 3718.

Cannibal Galaxies

NGC 5907

The galaxy NGC 5907 is an Sc galaxy seen within a few degrees of edge on. It lies at a distance of about 40–50 million light-years in the constellation Draco, the Dragon. It has a bright centre but a very small bulge. The dark lane at the edge of the disc can be clearly seen against the background of the galaxy.

The edges of the disc at each side show evidence of the open spiral arms of the galaxy, thus giving it the Sc classification that it has. On one side the spiral arms can be clearly seen pointing away from us, while on the other edge the outer spiral arms are pointing towards us.

Looking closely at the disc, we can see that the edge nearest to us does not exactly follow a straight line. In other words, the disc bends slightly. This makes it a galaxy with what is known as a warped disc.

There are no other sizeable galaxies in the vicinity, so the warp was not created by the gravitational effect of a passing large galaxy – which would be what is needed to create such a noticeable warp. Hence, it is often considered to be a prototypical example of an isolated warped disc galaxy. How these warps occur in galaxies is still a matter of debate among astronomers.

The most striking aspect of this picture is the large extended streamer surrounding the galaxy, almost as if it were tying the galaxy in a gigantic knot. This feature was not seen in early images of this galaxy and was only discovered in this century with the advent of modern detectors and specialist software.

What astronomers think has happened here is that a relatively small object, such as a dwarf elliptical galaxy, has fallen into NGC 5907. In doing so, it has passed straight through, but the effect of this is that the smaller galaxy has started to lose its stars by the process of tidal stripping.

Tidal Stripping

Tidal stripping is the name given to an effect that sometimes occurs during the interactions between galaxies. The gravitational attraction of one galaxy can pull out either the stars or the interstellar medium of the other galaxy, literally stripping the galaxy of its contents.

As the smaller galaxy carries on through the larger galaxy, it leaves behind a trail of stars, which are seen as a large, luminous ribbon. Elliptical galaxies generally have no interstellar medium, so this tidal stream consists only of stars. The ribbon is known as a 'star stream'.

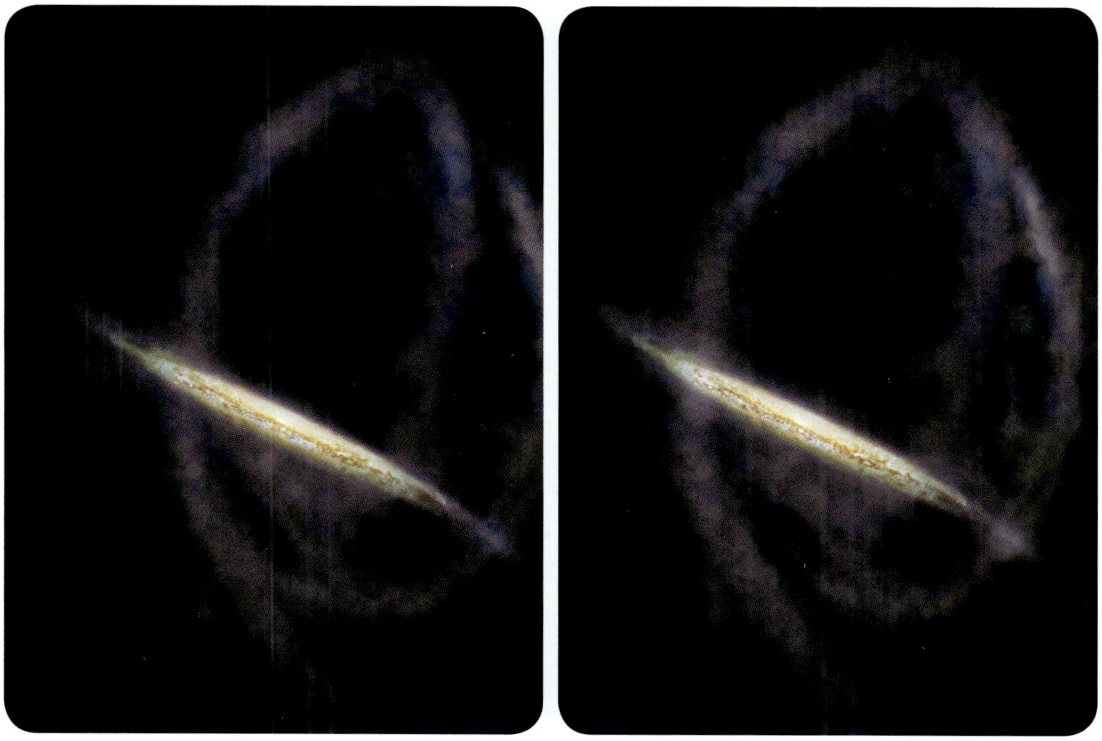

The galaxy NGC 5907, with large extended streamers appearing to tie it in a knot. These are known as star streams and are the result of a smaller galaxy passing through NGC 5907, leaving behind a trail of debris through a process known as tidal stripping. Note this is a real photograph and not an artist's impression.

The effect is not dissimilar to that seen when a comet passes through the inner Solar System and leaves behind it a trail of debris, which causes a meteor shower when Earth passes through the debris. Note that the mechanism that causes the trail is different for a comet, where it is usually known as a 'tail'. It is the result of heating from the Sun, whereas in galaxy interactions it is a gravitational effect, but the net result is similar.

The dwarf galaxy then appears not to have been able to escape from the gravitational attraction of NGC 5907, reaching a maximum distance of 150,000 light-years above the plane of the larger galaxy before falling back down into the bigger galaxy again. It seems to have done this several times on either side of NGC 5907 until there was nothing of it left other than a beautiful trail of star streams.

We do not have a name for this destroyed dwarf galaxy because all of this happened long before astronomers came along to give galaxies names. However, we can admire its legacy in the streams around NGC 5907.

The Milky Way and the Sagittarius Dwarf galaxy

It came as a surprise to many astronomers, as it will no doubt come as a surprise to some of our readers, to learn that our own Milky Way is an interacting galaxy, albeit with a very much smaller galaxy.

In the mid-1990s a team of astronomers from Cambridge published in *Nature* their discovery of a very nearby dwarf galaxy, in the constellation Sagittarius, which appeared to be interacting with the Milky Way Galaxy.

A subsequent flurry of activity by astronomers, using telescopes around the world, found a series of star streams similar to those we saw around NGC 5907. So, if observed from afar, our Milky Way Galaxy might look a little bit like NGC 5907.

> **Galactic Coordinates**
>
> When astronomers produce star charts, they plot the positions of celestial objects as if we were at the centre of a sphere to which these objects are attached – much as ancient civilisations thought was physically the case! Just as geographers assign coordinates of latitude and longitude to objects on the Earth's surface, in astronomy we can assign spherical coordinates to every star, nebula, galaxy and so on. It is often useful to define Galactic coordinates based on the plane of the Milky Way Galaxy. In this system, the horizontal coordinate is defined in Galactic longitude (l), and the distance 'above' or 'below' the Galactic plane is designated Galactic latitude (b).

Astronomers identified the star streams by using the fact that elliptical galaxies have little or no interstellar medium and therefore have no active star formation; hence all their stars are quite old. The galaxy interacting with the Milky Way Galaxy is known to be a dwarf elliptical galaxy and hence will contain only old stars. So astronomers targeted what are known as 'M dwarf stars', which are a type of old, low-mass star that is bright in the infrared.

So, by using an infrared survey, astronomers were able to find M dwarf stars at the right distances and plot their distribution on the sky. This is shown in the image (next page), where the darker a region is, the more M dwarf stars there are in that area.

We can see a number of bands or streams across this image. The most prominent starts at the left-hand side of the image at 50 degrees below the plane of the Galaxy, rises up to a maximum of about 50 degrees above the plane, and then back down again. Of course the right-hand side of the plot joins back onto the left-hand side of the image – it is continuous. There are two or three other bands that are somewhat fainter. Where they cross, they produce the darkest regions of the plot.

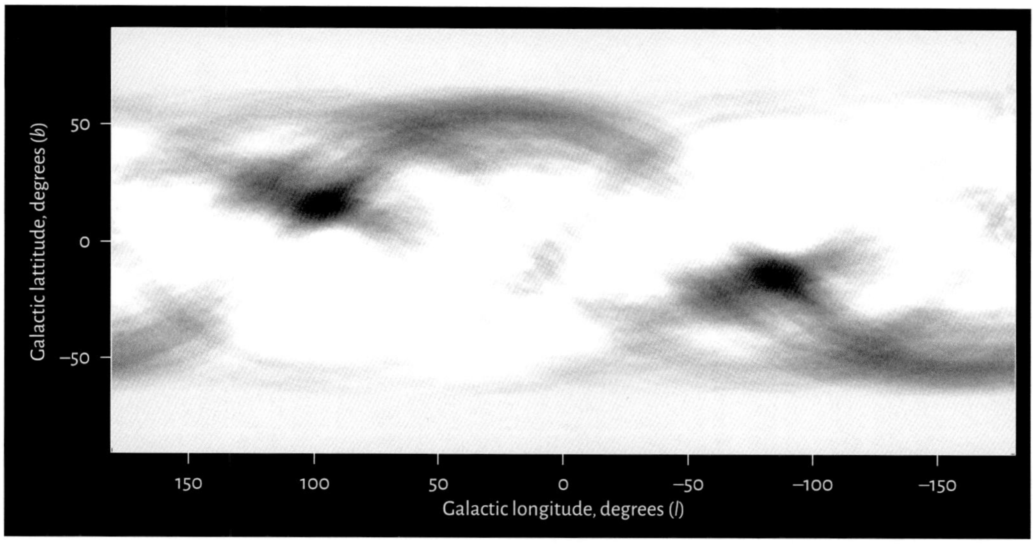

A real picture of the whole sky, shown with Galactic coordinates, of the distribution of stars believed to have been trailed in the star stream behind the Sagittarius Dwarf galaxy. The darker the region, the greater the number of stars.

These streams give astronomers a unique opportunity to study at close hand, relatively speaking, the inside of another galaxy. The Sagittarius dwarf galaxy has been torn apart by the gravity of the Milky Way and spread across the sky for astronomers to study. This is the process of galaxy dissection we introduced earlier.

Since this plot was made, astronomers have studied many more stars and many more types of stars. They have done this not only by taking pictures at different wavelengths, such as the near-infrared image on the next page, but also by studying the different types and ages of the stars that make up the star streams.

This allows astronomers to tell, for instance, if all the stars were born at the same time or if there was continuous formation of stars over a long period of time. Additionally, if a star contains only hydrogen and helium, the two elements created in the Big Bang at the start of the Universe, then it is likely to be a first-generation star, born not long after the Big Bang. However, if a star contains a substantial fraction of heavier elements, then it has to be a later generation star, born after the first massive stars turned into supernovae, drenching their surroundings in heavier elements.

Astronomers are still debating the exact details of their findings on this, but the evidence seems to suggest that the formation of stars continued in this dwarf galaxy for the first few billion years after the Big Bang. Thereafter, the number of new stars being born dropped sharply when the Universe was about half of its current age.

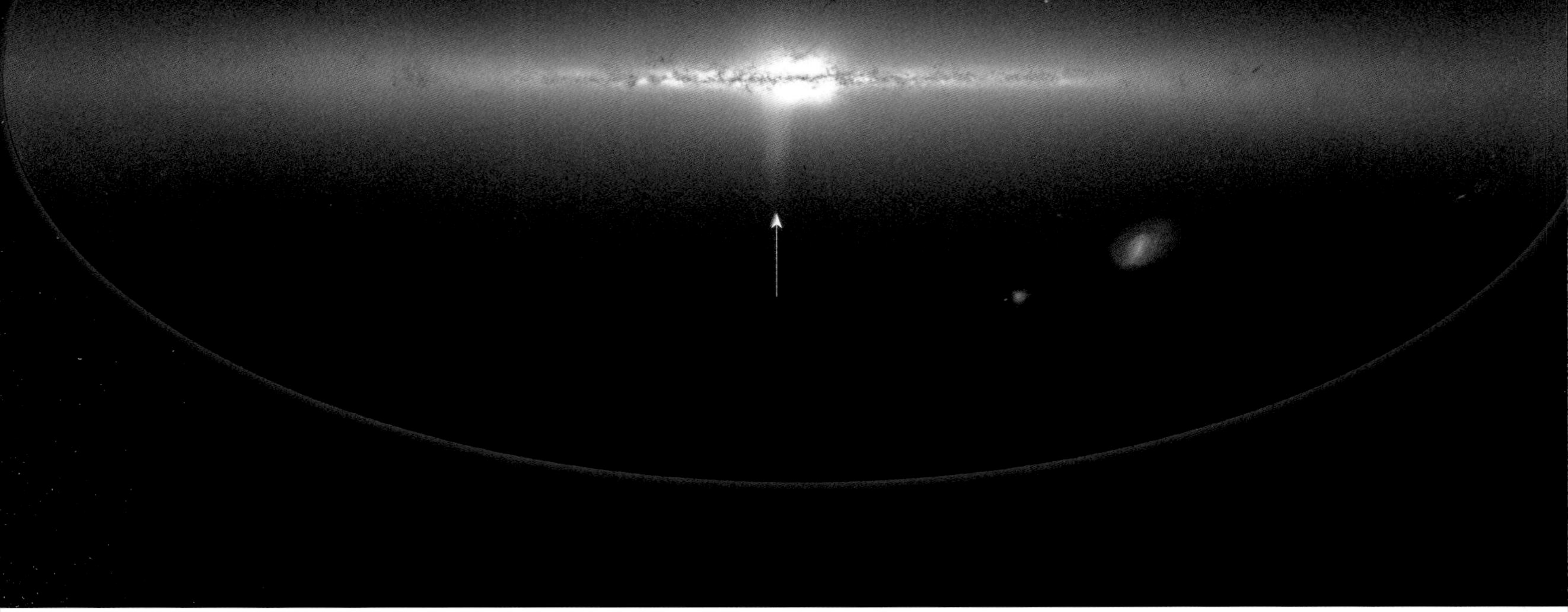

The Milky Way Galaxy taken in the near infrared. Note the faint vertical strip extending below the Galactic Centre, which is the start of the star stream (marked by an arrow).

This may have been because it used up its interstellar medium, and so had no matter with which to form new stars. It is also possible that its star formation stopped on first encounter with the Milky Way. Elliptical galaxies are known to have very little interstellar medium anyway, so it is possible that it was a combination of the two effects.

More recently, it is possible that there was another spurt of star formation in the last couple of billion years. This could have been triggered by a series of supernova explosions in the remnant of the dwarf galaxy, thereby replenishing its interstellar medium and allowing a new generation of stars to be born.

Astronomers can study plots, such as that shown on the previous page, together with the measured distances of all the stars, and produce a three-dimensional distribution of the star streams in our Galaxy that have been tidally stripped out of the Sagittarius Dwarf galaxy. This can be plotted in such images as the artist's impression on the next page to show what the star streams might look like from afar. Hence we see that our Milky Way Galaxy may indeed look rather similar to NGC 5907 (shown earlier in this chapter).

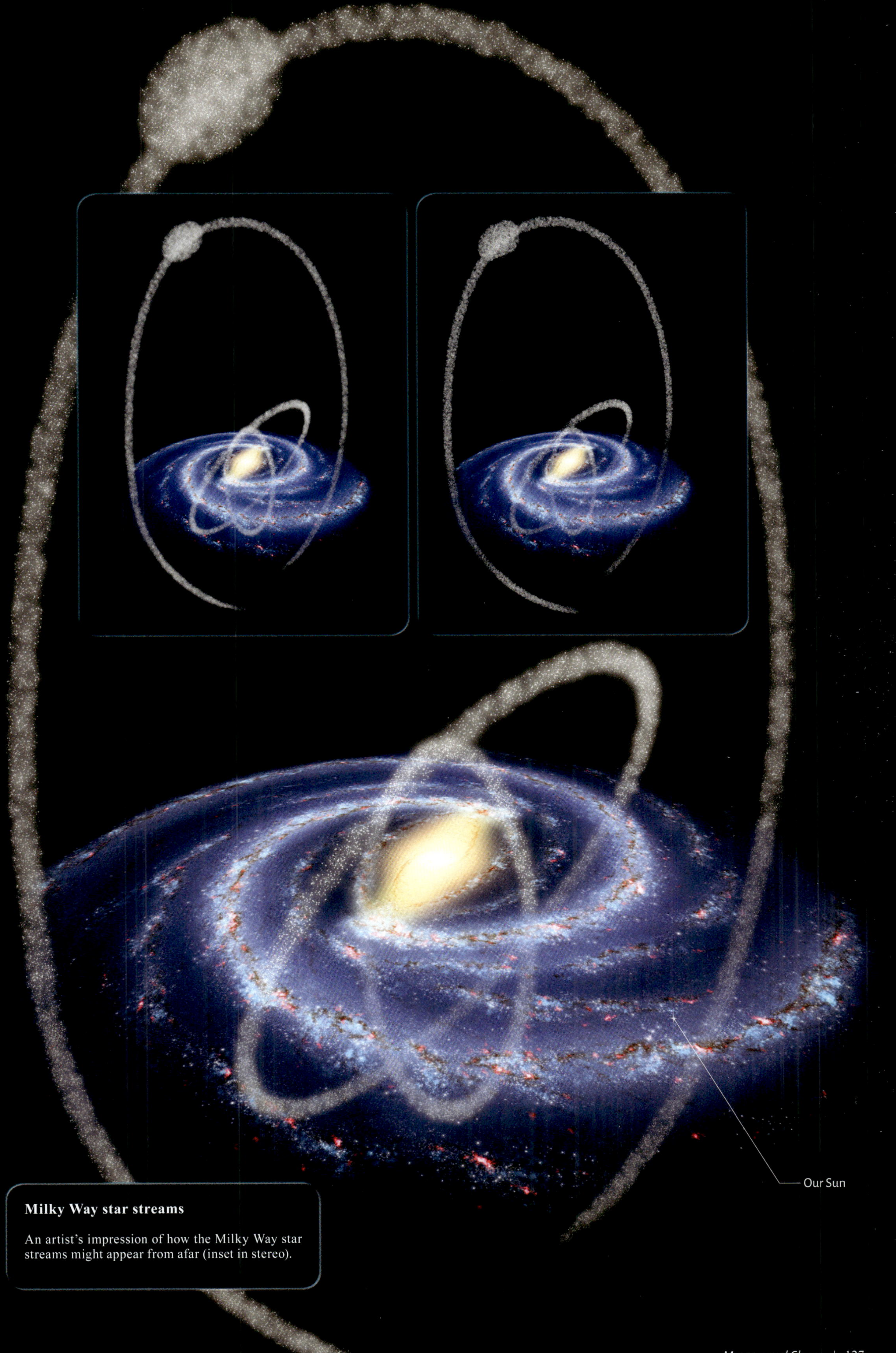

Milky Way star streams

An artist's impression of how the Milky Way star streams might appear from afar (inset in stereo).

Train Wrecks of the Cosmos

So far, we have looked at flybys and what might be regarded as gentle collisions between galaxies. We now look at what happens when two giant galaxies collide, whether spiral, elliptical or irregular.

We will choose collisions that illustrate different stages and phases of the colliding process. We start with galaxies that are approaching one another, followed by galaxies in the midst of tearing each other apart, and finish with galaxies in the process of separating.

In this way, we hope to show the equivalent to a time-lapse movie of a single collision, without having to wait a few billion years to watch it happen!

Early Stages

NGC 4567 and NGC 4568

The spiral galaxies NGC 4567 and NGC 4568 are a pair of colliding spiral galaxies in the constellation Virgo and are part of the Virgo Cluster (see Chapter 7). They lie at a distance of about 60 million light-years. Both galaxies are classified as Sbc - unbarred spiral galaxies with well-defined open spiral arm structure.

In the picture (next page), it looks as if NGC 4568 is going to cut through NGC 4567 like some gigantic circular saw, neatly cutting it in half. However, it does not work quite like that because, of course, galaxies are not solid bodies.

What is more likely to happen is that, as the galaxies approach one another, they will start to orbit each other in overlapping orbits. Their outer regions will entwine with one another and, as the galaxies continue their orbital dance, they will eventually either merge into one supermassive galaxy, or they will move apart again, leaving massive star streamers and giant tails behind them. Which one of these two scenarios occurs depends on the detailed dynamics of their collision.

The dark lanes of the upper, and apparently larger, galaxy NGC 4568 can clearly be seen to lie in front of the lower galaxy NGC 4567. Hence, we can say that NGC 4568 is closer to use than NGC 4567. However, there is still some debate among astronomers over whether they are genuinely interacting or just happen to lie along the same line of sight - a so-called chance alignment.

For example, *The Cambridge Photographic Atlas of Galaxies* states that the galaxies are consistent with being at the same distance, while admitting that there are large uncertainties on these measured distances. The authors argue that there are no signs of any interaction taking place between the two, and hence that they lie at different distances and just happen to lie along the same line of sight.

We would argue otherwise, since the probability of a chance alignment this close is very low, and because these two galaxies are clearly in the earliest stages of interacting, and so the amount of interaction would be expected to be very small.

The colliding galaxies NGC 4567 (below and behind) and NGC 4568 (right foreground), captured by the International Gemini Observatory, Hawaii. NGC 4568 looks as if it is about to cut its way through NGC 4567 like some gigantic cosmic circular saw.

Further, in the upper right-hand part of the image, we can see the dark lanes of NGC 4568, and they show signs of possibly having been pulled away from the rest of the galaxy. This may be the first sign of interaction between the two galaxies. We therefore suggest that this picture shows two galaxies approaching one another in the very earliest stages of interaction.

It is interesting to note that the image might indicate what an observer who was far enough away would see in about 5 billion years' time, when the Milky Way Galaxy and the Andromeda galaxy are expected to collide.

Mid-Collision

Arp-Madore 417-391

The interacting pair of galaxies known as Arp-Madore 417-391 lies in the southern hemisphere constellation Eridanus, the Celestial River. The galaxies are very distant, at over 600 million light-years away.

The pair presents a fascinating picture, appearing as a ring with the two galactic centres clearly visible side by side, but on one side of the ring. The remainder of the ring appears to be made up of a combination of the spiral arms of the two galaxies. The dark lanes from one of the spiral arms can still be seen on one side of the ring in the image (next page).

How the original galaxies came to end up in this most unusual configuration is a matter for speculation. We can hypothesize that the two galaxies went into orbit around one another and their spiral arms were pulled out into the shape of the orbit, in a more extreme example of the streams and tails we saw above. Presumably, after long enough, the remnants of the spiral arms filled the orbits, creating the ring that we see.

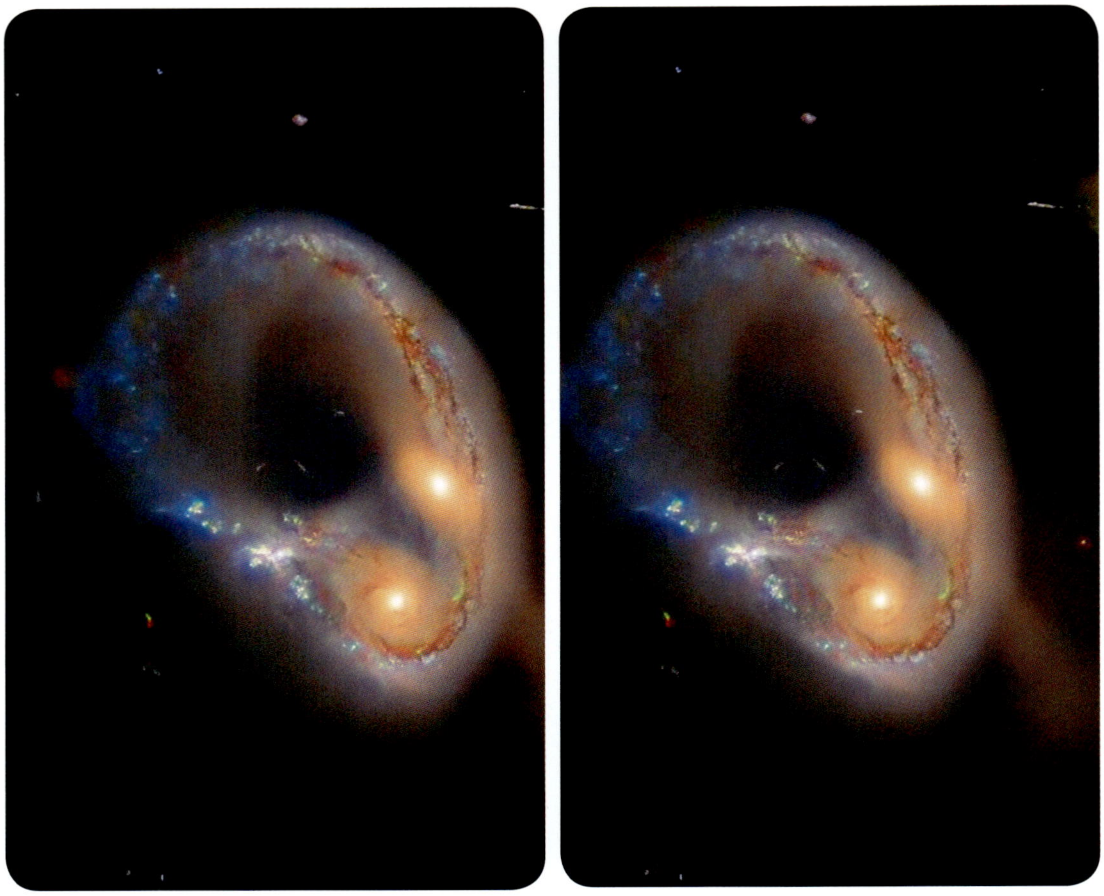

The colliding galaxies known as Arp-Madore 417-391. The centres of the original two galaxies can be seen lower right, and their spiral arms have been pulled into a ring upper left.

The future evolution of this pair is probably that their original galactic centres will merge and form one single, large galactic centre. What happens to the ring is less clear. It may be lost into space, in which case the final galaxy may end up as an elliptical galaxy, or it may collapse down onto the combined centres, or be dragged into a peculiar shape, leaving an irregular galaxy. Only time will tell.

NGC 2623

NGC 2623 lies at a distance of some 250 million light-years from us in the constellation Cancer. It appears to be two galaxies that are completely entwined in some kind of giant cosmic embrace.

We saw above how the galaxy NGC 3718 had its spiral arms pulled out straight after an encounter with a passing galaxy. Here, both galaxies have had their spiral arms pulled straight by the encounter with each other, creating giant tidal tails.

In the tidal tails we see evidence of the dark lanes that were previously strung out along the spiral arms. We also see small bright regions where stars are still forming.

The centre of the collision is extremely messy. However, it is still possible to see two bright centres, one reddish and one more bluish. The reddish one clearly contains a lot of dust because it partly obscures the bluish one, which must therefore be behind it.

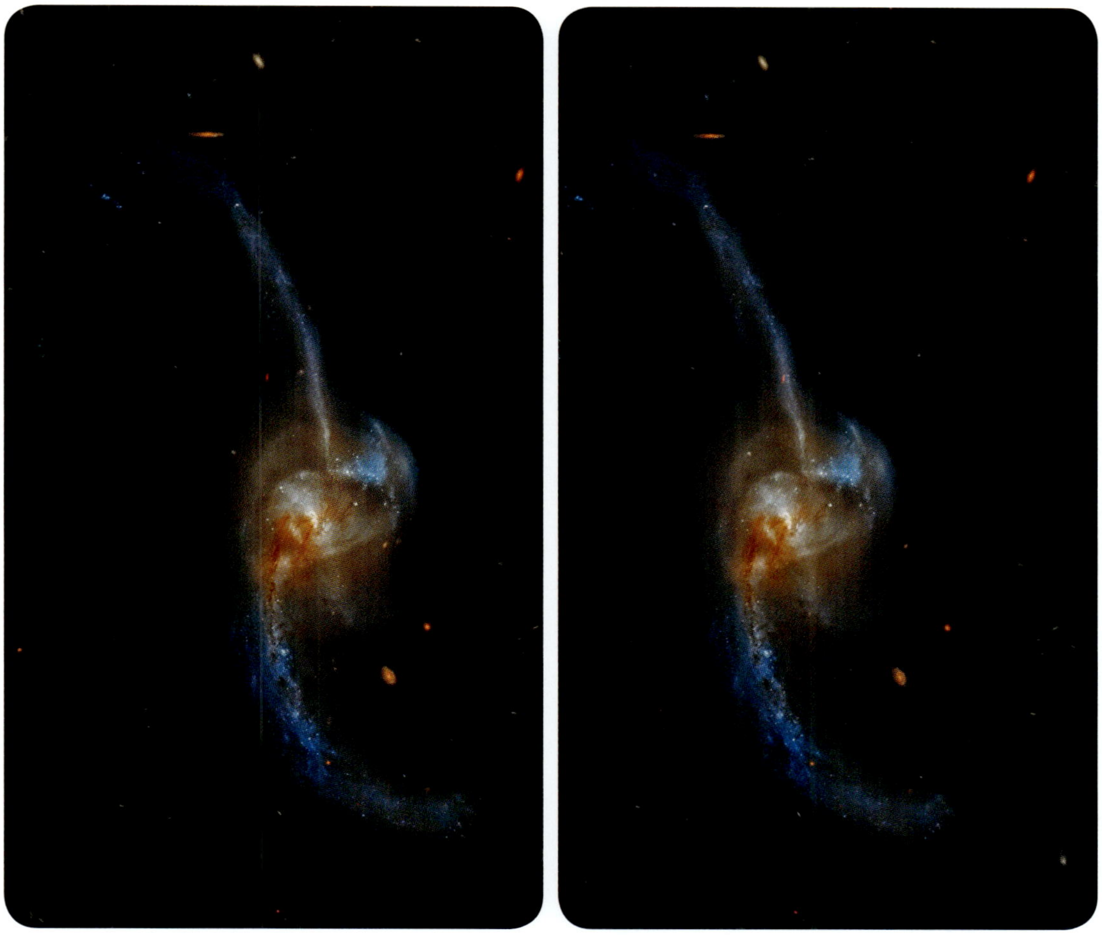

The colliding galaxies known as NGC 2623 showing highly developed tidal tails. The two galactic centres have merged, and the spiral arms have been dragged out into the tails.

Astronomers can only speculate as to what types of galaxies these two were before they collided. However, the small dark lane seen next to the background bluish galactic centre suggests that it may once have had a bar. The larger linear feature crossing the background galactic centre is probably caused by the collision itself.

Astronomers have also studied X-rays from these galaxies, and the pattern of emissions is similar to that seen in a starburst galaxy. For a discussion of starburst galaxies see M82 in Chapter 4, but basically such a galaxy has a large number of stars forming simultaneously, usually in the centre of a galaxy. In our picture, the galaxy with the reddish centre is most likely the starburst galaxy because the dust that causes the reddish colour is a sign of regions where stars are forming. Hence, by closely studying the collision itself, we can deduce a lot about the two galaxies before they collided.

NGC 4038/Caldwell 60 and NGC 4039/Caldwell 61 – the Antennae galaxies

NGC 4038 and NGC 4039 are two galaxies that can be seen in mid-collision. They are known by astronomers as the Antennae galaxies, and they are in many ways the poster children of colliding galaxies, often photographed and published. They lie in the constellation Corvus. They also appear in the Arp Catalogue as Arp 244.

They were probably both originally spiral galaxies. The two bright, yellowish white galactic centres can be seen towards the top and bottom of the image. A distinct spiral arm can be seen surrounding the upper galactic centre. It is picked out in blue by the bright stars in the arm and in red by the stars still forming.

The lower of the two galactic centres can be seen to be surrounded by a small, residual spiral arm. In between the two is a mass of gas and dust picked out by dark lanes and red patches of star formation that is the remnant of other spiral arms from the two galaxies. The large amount of star formation taking place leads astronomers to label this pair as starburst galaxies.

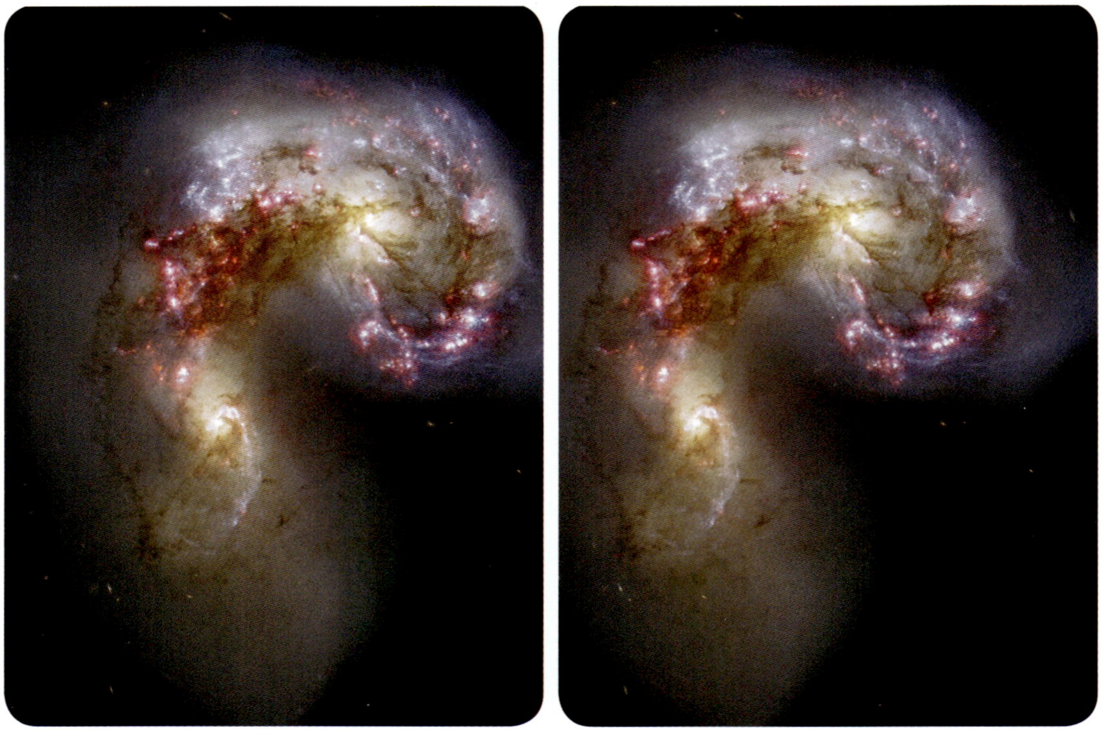

The colliding galaxies known as the Antennae galaxies provide one of the most beautiful pictures in all of astronomy.

Post-Collision and Oddities

Some galaxies look so unusual that it is hard to envisage what kind of collisions might have caused them, or what catastrophic fate they underwent to end up looking as they do. We now examine a few of these.

NGC 470 and NGC 474

The galaxies NGC 470 and NGC 474 lie in the constellation Pisces at a distance of about 100 million light-years. They are a pair of possibly interacting galaxies that do not appear to overlap in our picture. So assuming they are actually physically close together, as opposed to accidentally appearing in the same line of sight, they must be still approaching one another, undergoing a near miss, or separating or orbiting one another after a flyby. The complex structure that we see in the image of NGC 474 (the larger galaxy in our image) is quite striking and suggests that this galaxy has indeed undergone a very significant interaction.

However NGC 470, classified as a spiral galaxy of type Sb, with open spiral arms and a relatively small bulge, has an apparently undisturbed structure. This observation has led some astronomers to believe that these two galaxies are only weakly interacting with one another. If this is the case, then we are left with the question of how NGC 474 came to look as it does.

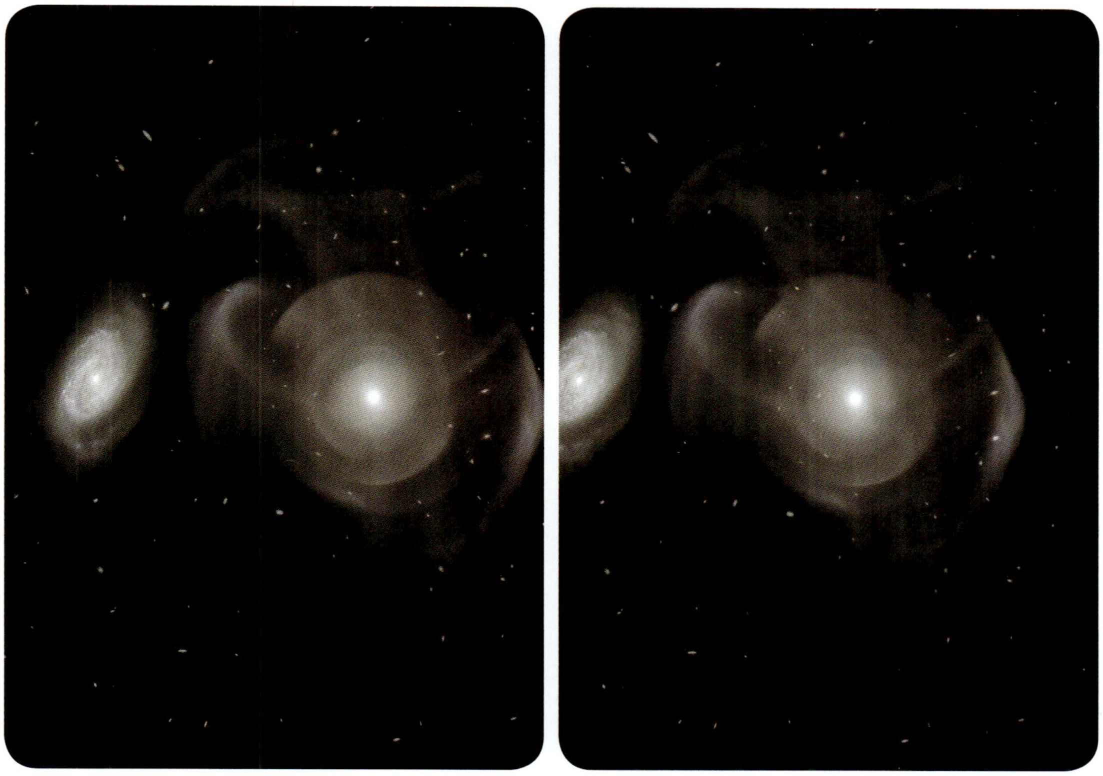

The galaxies NGC 470 (left in the stereo image) and NGC 474 (right).

NGC 474 is classified variously as a lenticular S0 galaxy, as an elliptical galaxy, or what is known as a 'shell' galaxy due to the concentric rings seen across its face. However, in the image above we also see an arc of matter in front of the centre of the galaxy, seemingly at right angles to the concentric rings.

The Keck Observatory is a pair of 10-metre telescopes located on top of Mauna Kea, an extinct volcano on the island of Hawaii; it was used to study NGC 474.

Recently, a team from the Keck Observatory studied NGC 474 and deduced a number of things about the galaxy. They found that it was consistent with having originally been either a lenticular or an elliptical galaxy that merged with another galaxy around 2 billion years ago, which it has essentially swallowed. This explains the concentric rings, or shells, which are rather like ripples on a pond after a stone has been thrown into it.

The team used the Keck 10-metre telescopes to take spectra of the central region of NGC 474, and also the outer shell regions. From this they learned that the centre of the galaxy is populated by stars that are roughly 8 billion years old, but the stars in the outer shells are only about 4 billion years old, similar to the age of the Sun.

They hypothesized that the old stars in the centre of the galaxy come from the original elliptical galaxy, whereas the stars in the outer shells come from the galaxy that merged with NGC 474. It would therefore be necessary to assume that the galaxy that merged into NGC 474 has been literally torn apart, and its stars scattered across the entire resulting shell galaxy.

They also found evidence that NGC 474 is pulling matter in from NGC 470, albeit at a relatively slow rate. The dust-rich and gas-rich matter in the spiral galaxy NGC 470 will replenish the gas-poor and dust-poor interstellar medium of the elliptical galaxy NGC 474. This should permit NGC 474 to restart its internal star formation in the future.

Astronomers remain baffled by the feature that crosses the face of NGC 474. Various suggestions have been put forward to explain it. One idea is that it may have been formed in a separate merger event. Another idea is that it is a streamer, such as we saw earlier in our own Galaxy and in NGC 5907, caused by a dwarf galaxy passing through the centre and dragging out a tidal tail. Astronomers continue to search for an answer.

NGC 2685

The galaxy NGC 2685 is also known as the Helix galaxy. It lies at a distance of roughly 50 million light-years and is in the constellation Ursa Major, the Great Bear (also known as the Saucepan or the Big Dipper).

Images of NGC 2685 show a bright, elongated central structure, believed to be a lenticular galaxy, or in the Hubble classification an S0 galaxy (see Chapter 4). This is a galaxy intermediate between a spiral and an elliptical. It contains an elliptical core surrounded by a featureless disc. Hence it has some of the attributes of both a disc galaxy and an elliptical galaxy.

NGC 2685 also has a series of rings that surround the central S0 galaxy. The surprising aspect of this is that the rings, which are clearly dusty, lie in a plane at right angles to the plane of the galaxy's disc. This form of structure is known as a 'polar ring' galaxy.

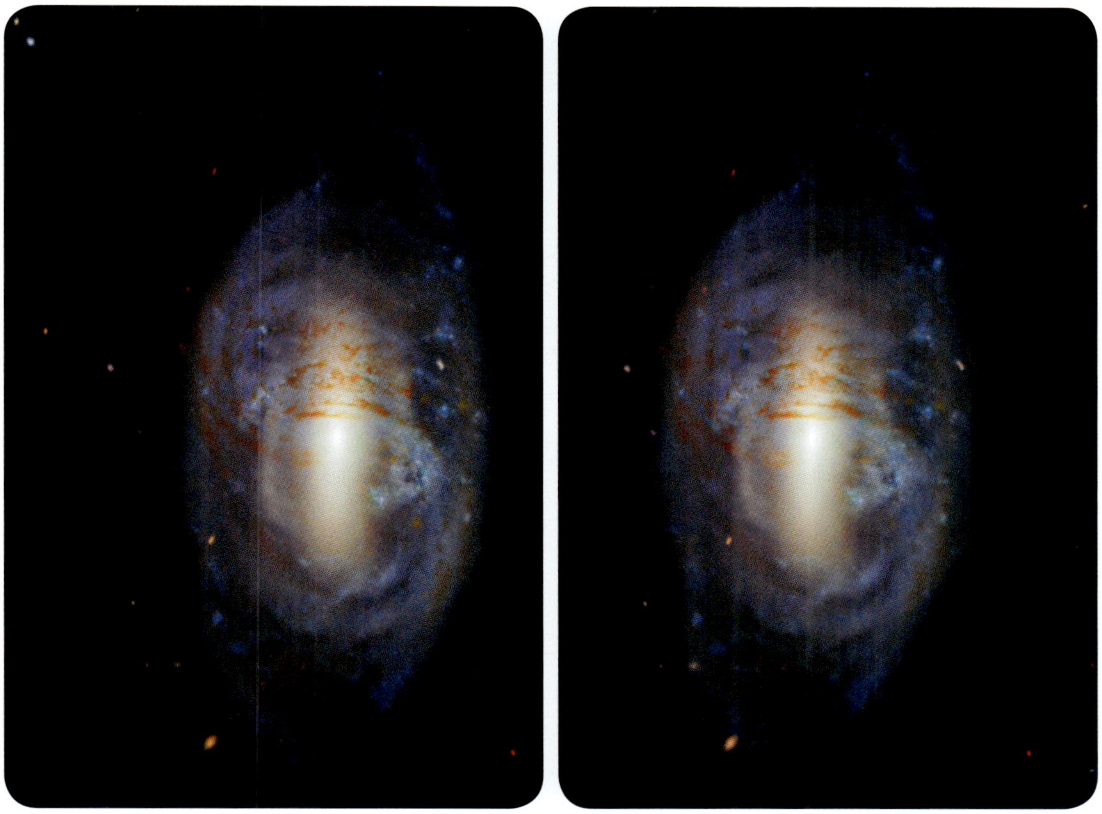

The polar ring galaxy NGC 2685 has an elongated structure together with a series of dark rings at right angles to the direction of elongation.

Polar ring galaxies are believed to form through a merger of two galaxies. The original central lenticular galaxy interacts with a smaller passing galaxy, which is travelling in a direction perpendicular to the disc plane of the lenticular galaxy. The smaller galaxy is captured by the gravitational field of the larger lenticular galaxy and goes into an orbit perpendicular to the disc of the lenticular galaxy.

During the course of several orbits, the smaller galaxy is gradually destroyed by the gravitational field of the larger galaxy, and leaves a series of rings in its wake. These are known as polar rings because they are in a polar orbit rather than an equatorial orbit around the disc. Polar ring galaxies were discovered in the 1970s.

With NGC 2685, the polar rings clearly contain dust as they can be seen in shadow against the central bright galaxy. Thus the incoming galaxy that was destroyed was originally a dusty galaxy - possibly a dwarf irregular galaxy or a small spiral galaxy.

Polar Ring Galaxies

Polar ring galaxies occur most commonly among the S0 lenticular galaxy population. The textbook *Galactic Astronomy* states that they are characterized as galaxies that have rings that pass over the poles of the central disc galaxy, rather than lying in the equatorial plane parallel to the disc.

A large number of lenticular galaxies have rings that are misaligned with the planes of their discs. *Galactic Astronomy* claims this could be as many as one-third. A much smaller number are genuinely polar ring galaxies. The *Cambridge Atlas of Galaxies* claims that this could be as few as 1 in 20.

The polar ring galaxy NGC 2685 from the Sloane Digital Sky Survey.

The renowned galactic astronomer Allan Sandage (1926–2010), who called NGC 2685 the 'Spindle' galaxy because of its long, thin shape with rings wrapped round it like string or thread, as it appeared in early images.

The 2.5-metre telescope that performed the Sloane Digital Sky Survey, which has now imaged more than 900,000 galaxies, including NGC 2685. The telescope is based at Apache Point Observatory in New Mexico, USA, and has an unusual design in that the entire building that houses the telescope slides off, leaving the telescope outside and free from the adverse observing effects of being inside a dome or other such building.

NGC 5128/Arp 153/Caldwell 77 – Centaurus A

The galaxy NGC 5128 is better known as Centaurus A, or Cen A. It is also sometimes known as Arp 153 because of its peculiar appearance. It lies at a distance of about 17 million light-years in the constellation Centaurus. It has been classified as either E7 or S0.

Centaurus A is reminiscent of the polar ring galaxies that we have just seen, such as NGC 2685. It may be the aftermath of a galaxy collision, where the dark area stretching across the face of the galaxy may simply be a dark lane rather than a ring.

When we look closely at our visible-light stereoscopic images, we do not see a ring, but rather an S-shaped dark lane stretching across the face of the galaxy. This could be the remnants of a spiral galaxy that has collided with the main part of the elliptical galaxy that forms the core of Cen A.

We can observe Cen A at other wavelengths to gain further insight into its nature and makeup. At X-ray and microwave wavelengths, it looks somewhat different. The foreground dark lane is still visible, but the elliptical body of the galaxy is no longer so prominent.

Furthermore, two bright jets can now be seen emerging in opposite directions from the centre of the galaxy, similar to the jets we saw in Chapter 2. These jets are indicators of a black hole at the centre of a galaxy. Hence we know that Cen A must contain a black hole at its centre. Whether this black hole was initially associated with the spiral galaxy or the elliptical galaxy before their collision is unknown.

However, we note that the jets lie perpendicular to the dark lane. In spiral galaxies with jets, the jets usually lie perpendicular to the plane of the disc - the plane of the dark lanes. Hence we can hypothesize that the jets and the black hole were all originally associated with the spiral galaxy before the collision.

There is also an entirely different explanation for the shape of Cen A. In our visible light images, it looks not dissimilar to some of the bulge galaxies that we saw in Chapter 5. It could simply be an edge-on galaxy in which the bulge is dominant, such as we saw in M104. However, this is not a widely held view.

Most astronomers argue that it is either a lenticular galaxy or a giant elliptical galaxy. In the middle of the 20th century, astronomers Walter Baade and Rudolph Minkowski speculated as to whether it was the result of a merger between a small spiral and a giant elliptical galaxy. Whatever its classification, it is a very interesting object.

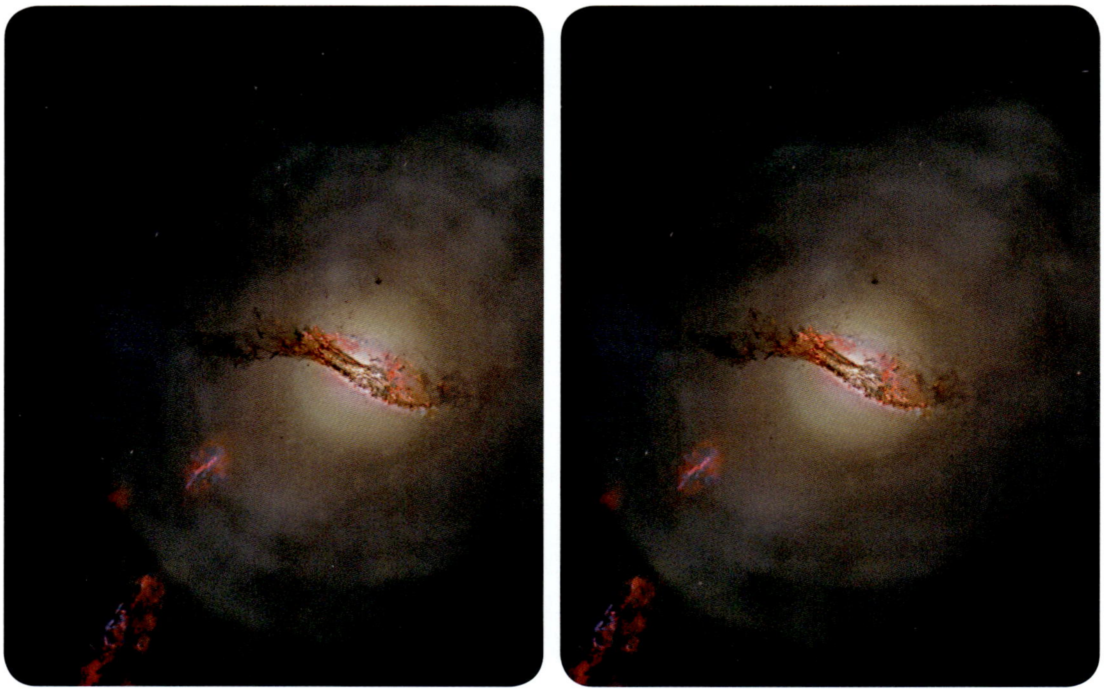

The galaxy Centaurus A seen in visible light.

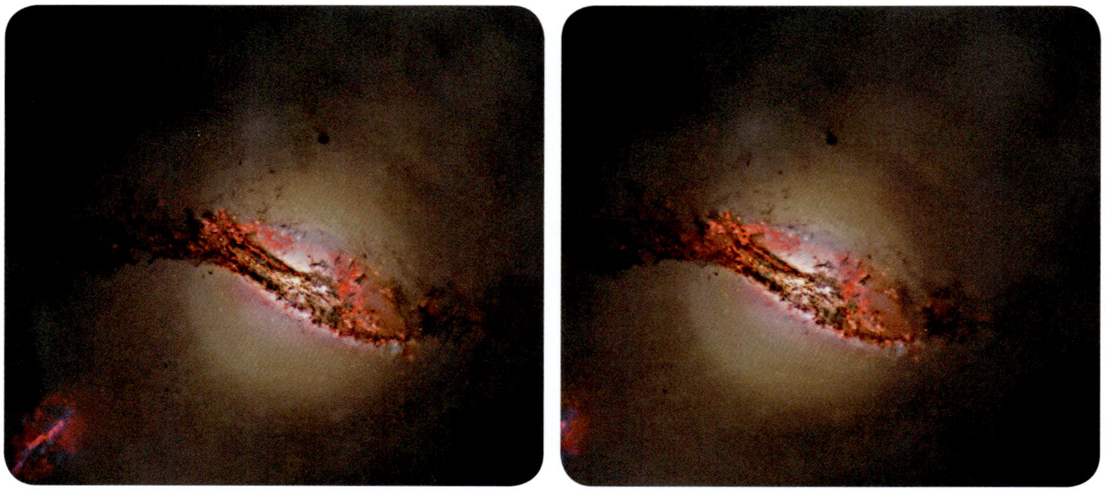

Detail of the centre of the galaxy Centaurus A.

The galaxy Centaurus A seen in X-ray and microwave wavelengths, flipped left to right.

NGC 2936/Arp 142 – the Porpoise galaxy, and NGC 2937

The Porpoise galaxy, NGC 2936, here imaged by the Hubble Space Telescope, is an interacting spiral galaxy in the constellation Hydra, at a distance of about 300 million light-years. NGC 2937, which can be seen immediately below it, is an elliptical galaxy, probably of type E3. NGC 2937 appears to have completely distorted NGC 2936, which is nicknamed the Porpoise because it resembles a porpoise leaping out of the water.

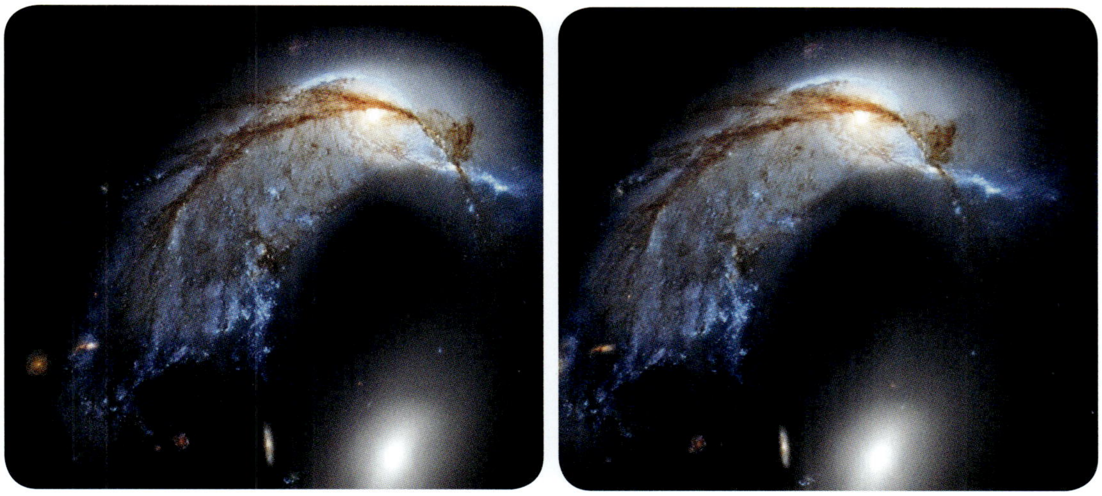

NGC 2936, the Porpoise galaxy and NGC 2937. The Porpoise was once a disc galaxy and has been torn apart by the gravitational influence of the elliptical galaxy NGC 2937.

NGC 2936 appears to have originally been a disc galaxy. The bright centre of the galaxy can be seen towards the top of the picture. The central bright region is intersected by a foreground dark lane. To the right of the galaxy's bright centre, a small stub of what was probably once a spiral arm can still be seen. This appears to indicate the direction of travel of the Porpoise during the interaction, in the same way that a comet leaves a trail behind it after interacting with the Sun.

To the left of the bright centre of NGC 2936, a trail of bright streamers extends behind the direction of travel. These streamers are what remain of the spiral arms of the original disc galaxy. The bright blue points of light scattered along the residual spiral arms show massive bright blue stars which are currently forming.

The dark lane that probably once bisected the original edge-on disc galaxy can still be seen curving along with the remnant of the disc. Many spiral galaxies that are seen edge on have a dark lane in the plane of the disc (see Chapter 5). This galaxy shows what happens when such a galaxy is distorted by the gravity of another nearby massive galaxy.

Looking closely at the bright yellow central region of NGC 2936, there is possibly evidence for a bar, or more likely a bulge. The spiral arms still start at either end of the bulge. Consequently, we can guess that this galaxy may have initially been an Sb or an Sc galaxy.

In the stereoscopic image, the dark lane can clearly be seen to be in front of the brighter parts of the galaxy. The dark lane probably continues to define the orientation of what remains of the plane of the galaxy. The galaxy streamers that were once the spiral arms have been pulled out to one side and below the plane.

The Aftermath: Ring Galaxies

One type of galaxy, known as a 'ring' galaxy, arises as a result of a specific kind of collision in which one galaxy passes straight through the centre of another (usually a disc galaxy originally). The latter then explodes outwards, creating the ring that gives it its name.

Arp 148 – Mayall's Object

The galaxy Arp 148 is also known as Mayall's Object, after its discoverer. Compared to most of the galaxies we have studied so far, this galaxy is at a very great distance, nearly 500 million light-years away in the constellation Ursa Major.

One of the earliest studies of Arp 148 was carried out by Margaret Burbidge in the 1960s, in which she worked out the details of how this peculiar object came to look as it currently does. What she surmised, and subsequent studies have since confirmed, is that Arp 148 is the aftermath of a collision between two galaxies. It looks as if one galaxy has passed right through another.

The images show what looks like a needle-shaped galaxy pointing at the centre of a giant blue disc with a broken-up or patchy appearance. The needle-shaped component may have originally been an extreme elliptical galaxy, or a lenticular galaxy seen edge on, with a dark lane across its middle.

The galaxy Arp 148. It looks like a needle passing through a disc.

Margaret Burbidge (1919–2020)

(Eleanor) Margaret Burbidge was a prominent 20th-century astronomer who was born in the UK and moved to the USA to continue her research career. She was one of the first astronomers to study Arp 148 in detail. However, she is better remembered for being part of a four-person team that worked out that almost all chemical elements were formed in the centres of stars. She was the first female President of the American Astronomical Society. *The Guardian* newspaper, in its obituary of her, acknowledged that she had challenged all forms of sexism in astronomy.

It is believed that the needle-shaped galaxy has passed straight through the other galaxy, which is of unknown original type. This resulted in the complete destruction of the second galaxy in a cataclysmic galaxy-wide explosion.

It is impossible to tell what type of galaxy the ring was before the needle-like galaxy passed through. However, what we can see at the centre of the circle, or slightly offset from the centre, is a bright yellow point of light that may be the remnants of the original galactic centre.

Margaret Burbidge was one of the first astronomers to study Arp 148 in detail.

This yellow centre contrasts with the blue of the rest of the ring. We can imagine that the ring may have been a spiral galaxy with an active yellow centre surrounded by blue spiral arms that were destroyed by the force of the impact.

The fairly symmetrical ring around the collision point arose from shocks coming outward from the centre, like a vastly more explosive version of the ripples emanating from a stone being thrown into a pond – in this case the stone has destroyed the pond!

The offset of the original galactic centre from the passage of the needle-shaped galaxy has come to be taken as evidence that the collision was off-centre. It has further been noted that the 'tail' of the needle appears to curve, upwards in the image, as it emerges from the last of the collision.

The Cambridge Photographic Atlas of Galaxies suggests that the collision was also affected by the gravity of a third galaxy, which is beyond the edge of our images but can be seen in others. The suggestion is that this third galaxy skewed the collision so that the needle-shaped galaxy did not pass through exactly in a straight line.

The ring galaxy does not appear to be perpendicular to the needle-shaped galaxy. Indeed, it appears to lean to the right at the top of the image. This adds weight to the hypothesis that the collision was not symmetrical.

There are many such examples of ring galaxies being formed by collisions between galaxies. Some leave partial rings and arcs, others leave complete circles, but none is more puzzling than the ring galaxy that we study next.

Hoag's Object

Hoag's Object is a ring galaxy, named after its discoverer, found at a distance of about 600 million light-years in the constellation Serpens. Its appearance is hard to explain.

We see a blue ring-shaped or toroidal structure stretching to an outer diameter of about 120,000 light-years. The ring is mostly blue, but with some red spots within it where stars are probably forming today.

In the centre of the ring there is a bright yellow circle about 17,000 light-years across. Just to the left of the yellow circle there is another background ring galaxy, purely by chance along the same line of sight. The central yellow circle is a spheroidal galaxy.

We know that ring galaxies occur when one galaxy passes straight through the middle of another galaxy, creating an explosive event and sending circular shock waves through the galaxy that was hit, thereby destroying its original structure, and leaving only rings behind.

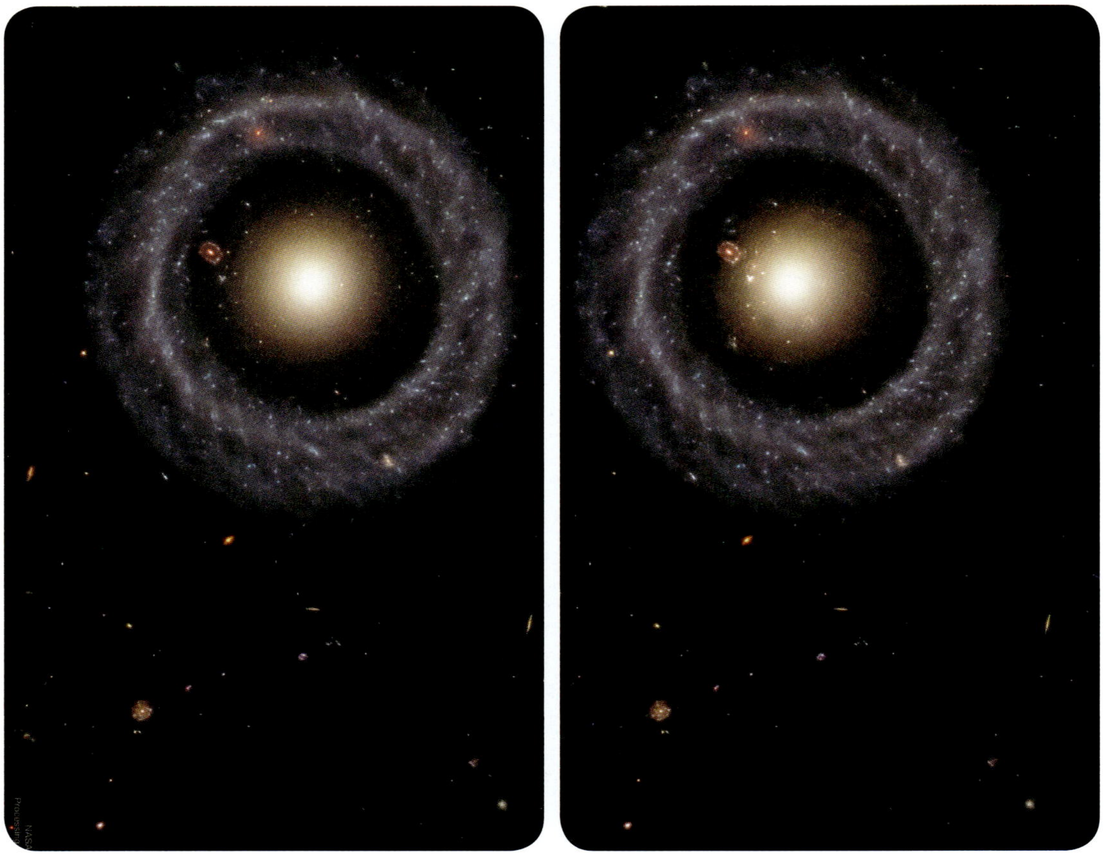

Hoag's Object looks like the perfect ring galaxy. This might be how Arp 148 would appear if we were looking at the needle end on.

The apparently obvious explanation for Hoag's Object is that the yellow circle is the galaxy that caused the explosion and created the ring. They are both at the same distance, so at first sight this seems plausible. But the problem with this idea is that both galaxies are moving at about the same speed; to create an explosion powerful enough to destroy a whole galaxy, the inner galaxy has to be moving at high speed relative to the outer galaxy. So this explanation simply does not work.

If another galaxy has previously passed through more quickly, then nobody has ever found it. There is no definitive explanation on which astronomers agree, so we leave this one as a mystery.

Centaurus A seen in optical wavelengths.

Stephan's Quintet

Stephan's Quintet (HCG 92) is a typical compact galaxy group. This large mosaic of images captured by the James Webb Space Telescope's Near-Infrared Camera (NIRCam) and Mid-Infrared Instrument (MIRI) covers the equivalent of about one-fifth of the Moon's diameter. It contains over 150 million pixels and is constructed from almost 1,000 separate image files.

To the Edge of the Universe

Clusters, Superclusters and Their Place in the Universe

Galaxies are the largest objects we know of in the Universe, but they are not the largest structures. As we have seen, galaxies coexist alongside other galaxies. They tend to appear in small groups, larger clusters and clusters of clusters, which we call 'superclusters'. We look at all these ways in which galaxies group together.

In this chapter we begin with compact groups and small groups. We then move on to groups, clusters (which are much larger than groups) and superclusters. We encounter filaments, walls and voids. We describe how astronomers have endeavoured to explain them in terms of gravity and dark matter. The simulations that astronomers currently run on supercomputers can do a good job of reproducing what we see in the Universe around us.

We finish with the very latest discoveries of enormous structures that astronomers cannot yet explain and are still debating. We take the reader to the very limits of what astronomers know about the roles that galaxies play in the Universe as a whole.

Galaxy Groups

Galaxy groups are collections of galaxies that are close to each other in space. The smallest are called 'compact groups' and typically contain only 5–7 galaxies. We look at two of these compact groups here. We also have small galaxy groups containing about 10–12 galaxies. The largest galaxy groups are generally defined as having up to about 50 galaxies – beyond which we call them 'clusters'.

Stephan's Quintet

Stephan's Quintet was named after its discoverer in 1877, and comprises the five galaxies shown on the previous page. These are NGC 7317, 7318A, 7318B, 7319 and 7320. The first four are actually members of a compact galaxy group, lying at a distance of about 250–300 million light-years. The fifth one, NGC 7320 (on the left of the picture), was later found to be a foreground galaxy, at a distance of about 40 million light-years, that just happens to lie along the same line of sight.

Hickson Compact Group 5 (HCG 5)

In 1982 astronomer Paul Hickson, from the University of British Columbia, produced a catalogue of about 100 compact galaxy groups. He searched the Palomar Observatory Sky Survey plates for groups of galaxies close together

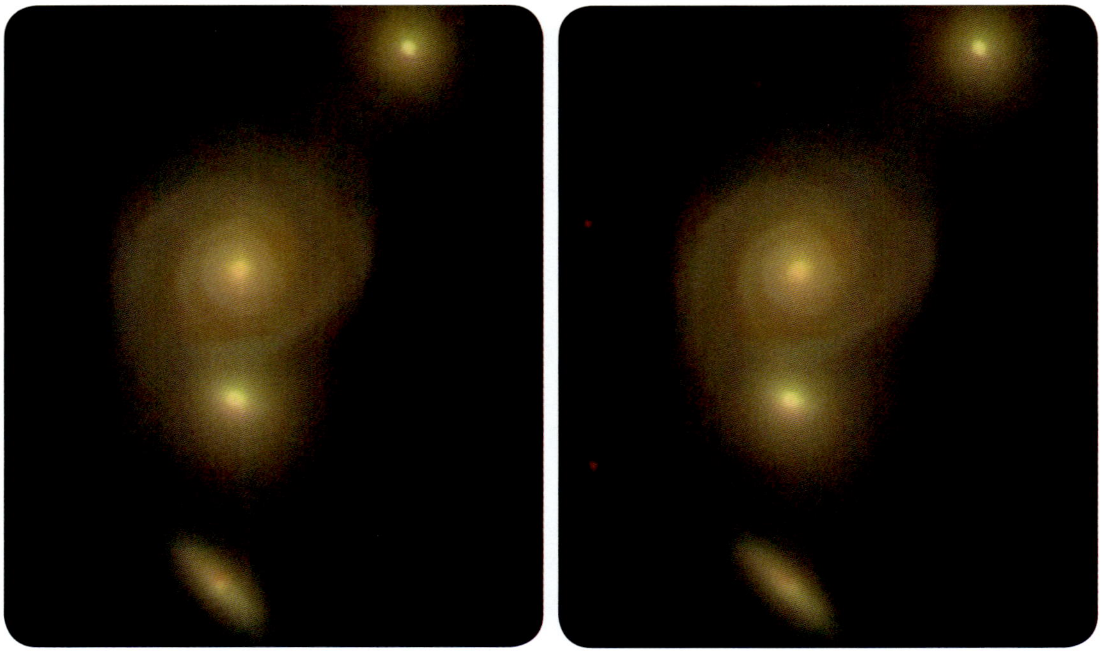

Hickson Compact Group 5 (HCG 5) is a typical compact galaxy group in the constellation Pisces.

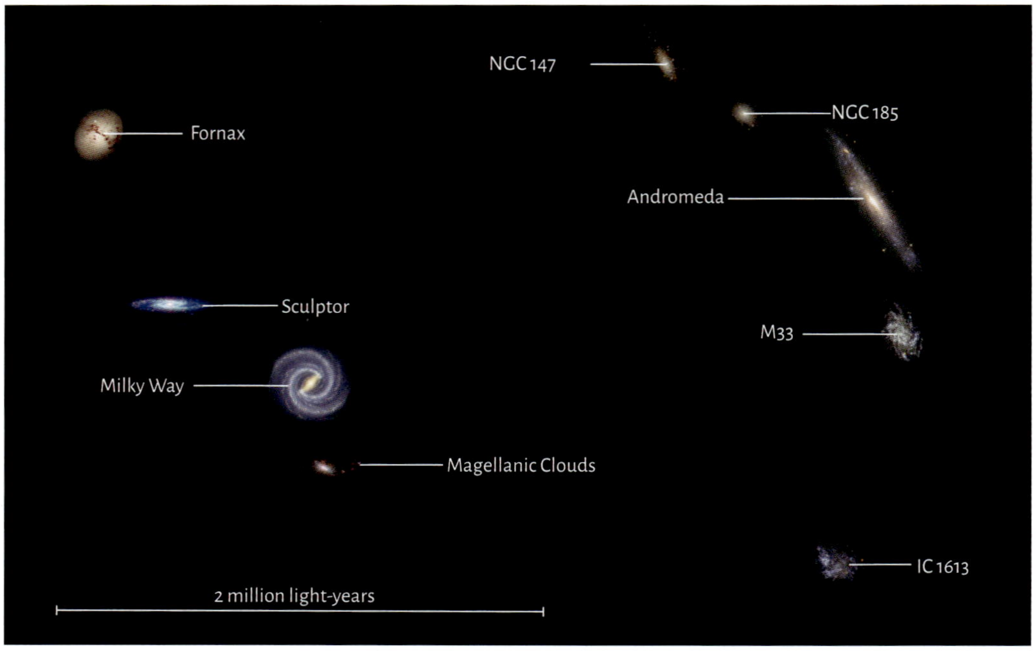

The Local Group of galaxies, showing the relation between the Milky Way, the Andromeda galaxy and other nearby galaxies.

and catalogued them, finding 100 groups. His main criterion in defining a group was that it must contain at least four, and not more than seven, galaxies of similar brightness that were isolated from other similarly bright galaxies, thereby excluding larger clusters. We have just looked at HCG 92, but we referred to it by its original name of Stephan's Quintet. We here look at HCG 5.

HCG 5 contains four galaxies. The two in the centre of the image are collectively known as NGC 190, a colliding pair of galaxies – an Sb open spiral galaxy above an elliptical galaxy of type E3. The one at the bottom is an edge-on spiral or lenticular galaxy, and the one at the top is an E0 elliptical galaxy. The group lies in the direction of the constellation Pisces, near the celestial equator, at a distance of about 550 million light-years.

The Local Group – a small galaxy group

The Local Group of galaxies is a typical small galaxy group, of which the Milky Way is a part. It was discussed in Chapter 4, where there are images of the galaxies, but we return to it again here, recapping the main points about its structure in order to set the context for what comes next.

The Local Group is dominated by two large spiral galaxies, the Milky Way and the Andromeda galaxy (or M31). Astronomers believe that in about 5 billion years these two galaxies will collide, but presently they are about 2 million light-years apart. Our Milky Way has two small satellite galaxies that we refer to as the Magellanic Clouds. The Andromeda galaxy also has a small satellite galaxy known as M110, and a slightly larger elliptical galaxy nearby that we call M32. The remainder of the Local Group consists of about half a dozen small galaxies that are mostly dwarf irregular, dwarf elliptical or spheroidal galaxies.

The physical observation that makes us call the Local Group an actual entity, rather than just a few galaxies that happen to inhabit the same part of the Universe, is that it appears that all of its galaxies are moving in roughly the same direction at a few hundred kilometres per second. This observation means that the Local Group galaxies are somehow tied together, and it is believed that they are bound by gravity. We refer to a situation such as this as a 'self-gravitating' group of galaxies.

The direction in which the Local Group appears to be travelling is towards the Virgo Cluster, which is the nearest cluster to us. It is much larger than the Local Group, containing many more galaxies, which are all once again bound together by gravity. The fact that the Local Group is travelling towards the Virgo Cluster implies that it must be gravitationally bound to the Virgo Cluster and hence qualifies as an outer part of the cluster.

Clusters of Galaxies

The Virgo Cluster

The centre of the Virgo Cluster is situated about 50 million light-years away from us in the direction of the constellation Virgo. It consists of about 2,000 galaxies and stretches for about 15 degrees across our sky (for comparison, the Moon is only a half a degree across). This means that it extends over 10 million light-years across the Universe. The Virgo Cluster is a truly gigantic entity. The total mass of the cluster could be as much as a million billion (a one with 15 zeros after it) times the mass of our Sun. It completely dominates our local Universe within a distance of 100 million light-years, as can be seen from the map of the local Universe on the next page. We now turn to examine some of the galaxies in the Virgo Cluster.

M87/NGC 4486

The galaxy M87, or NGC 4486, is believed to be the most massive galaxy in the Virgo Cluster. It is considered to be a supergiant elliptical galaxy due to its size and is of type E0. It lies close to, but not exactly at, the centre of the Virgo Cluster. It is the central dominant (cD, see box next page) galaxy of the Virgo Cluster.

Central Dominant Galaxies (cD)

cD galaxies typically lie at the centres of galaxy clusters. As the name suggests, they dominate the gravity and dynamics of the cluster. They are believed to grow by way of the mergers of a number of other galaxies that fall into the centre of the cluster. In essence, the cD galaxy consumes them, as we saw happening in individual cases in Chapter 6.

Like all cD galaxies, M87 is believed to have grown from the mergers of a large number of galaxies that fell towards the centre of the cluster under gravity. Astronomers believe that M87 may have consumed up to 50 normal spiral galaxies in this process! This would then explain why it is about 50 times more massive than the Milky Way.

We discussed M87 in Chapter 2 in the context of the black hole at its centre, which is 6 billion times more massive than the Sun. The galaxy also has huge jets of high-energy, highly charged gas shooting out from its centre that originate close to the black hole. They can be seen in images and are also detected at radio wavelengths.

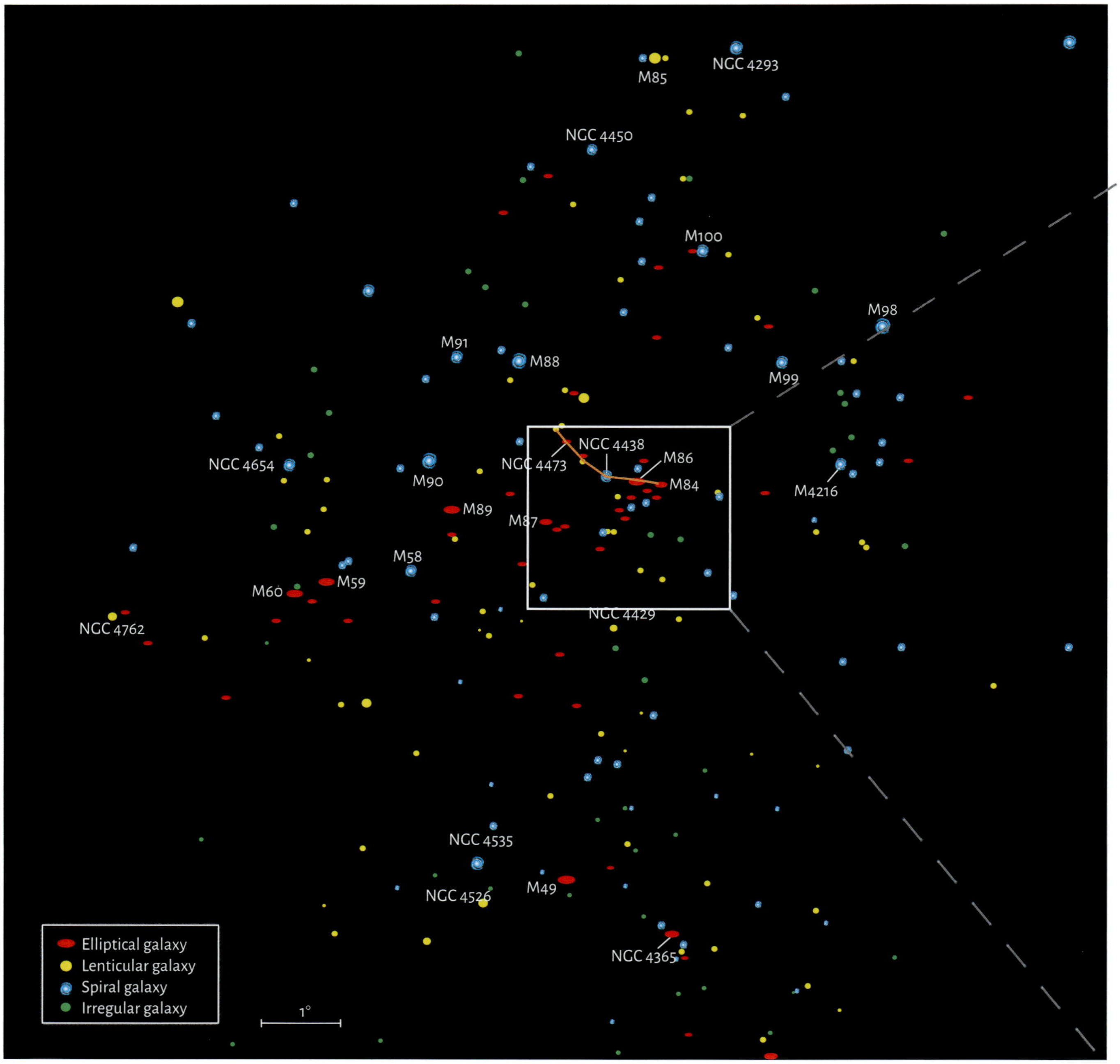

The local Universe within 100 million light-years is dominated by the Virgo Cluster of galaxies, shown here. The orange line indicates the apparently linked galaxies known as Markarian's Chain.

The giant elliptical galaxy Messier 87, or NGC 4486, at the heart of the Virgo Cluster, in which it is the most massive galaxy and is the central dominant (cD) galaxy.

A closer view of the Virgo Cluster

The Virgo Cluster showing detail of the region that contains Markarian's Chain (M84, M86, NGC 4435/38, NGC 4458/61, NGC 4473/77.

M49/NGC 4472

The galaxy M49, or NGC 4472, is another of the most massive galaxies in the Virgo Cluster. It is also a giant elliptical galaxy, but of type E2. It is referred to as a 'giant galaxy' because of its size: it is about 150,000 light-years across – more than twice the diameter of the Milky Way. It can be seen to have a halo, and it has roughly 5,000 globular clusters around it. It is believed to harbour a black hole at its centre that is roughly half a billion times the mass of the Sun. This is estimated from the discovery of a bright X-ray source at the galaxy's centre.

The giant elliptical galaxy Messier 49, or NGC 4472, is one of the most massive galaxies in the Virgo Cluster.

NGC 4216

The galaxy NGC 4216 is one of the brightest spiral galaxies in the Virgo Cluster. It lies at a distance of about 50 million light-years. It is a massive spiral galaxy and is intrinsically brighter than the Andromeda galaxy, even though it appears fainter because it is much further away.

It is seen very close to edge on to us, so its structure is difficult to discern. However, we can observe that it has a bright, white central core, surrounded by a yellow ring, which is brightest on the near side. In front of that can be seen well-delineated dark lanes, which are probably tracing the spiral arms. Hence it is probably a relatively open spiral of type Sb. Some astronomers believe it has a bar, but we maintain that it is too close to edge on for any bar to be discernible, so we are sticking with Sb.

The spiral galaxy NGC 4216 is one of the brightest spiral galaxies in the Virgo Cluster.

M98/NGC 4192

The galaxy M98, or NGC 4192, is a barred spiral galaxy classified as SBb or SBab, indicating the fairly tightly wound nature of its spiral arms. The bar is visible across the face of the galaxy, despite its high inclination angle of 75 degrees (that is, only 15 degrees from being completely edge on), as can be seen in the picture below.

It lies at a distance of about 40 million light-years from us, in the northern part of the Virgo Cluster and in the direction of the constellation Coma Berenices. It may have had a previous interaction with a galaxy nearby.

The spiral galaxy Messier 98, or NGC 4192, is one of the spiral galaxies in the Virgo Cluster.

Other Clusters of Galaxies

There are other associations of galaxies in our locality of the Universe, as can be seen from the finder chart below. In the constellation Dorado, for instance, there is an association of galaxies that is generally referred to by astronomers as a group. It is one of the largest galaxy groups that is not classified as a cluster.

Only slightly larger than the Dorado Group, and only slightly further away, are the Fornax and Eridanus clusters of galaxies, named after the constellations in which they are found. There are also much larger clusters that are much further away. We will start with a nearby cluster and then look at one that is more distant.

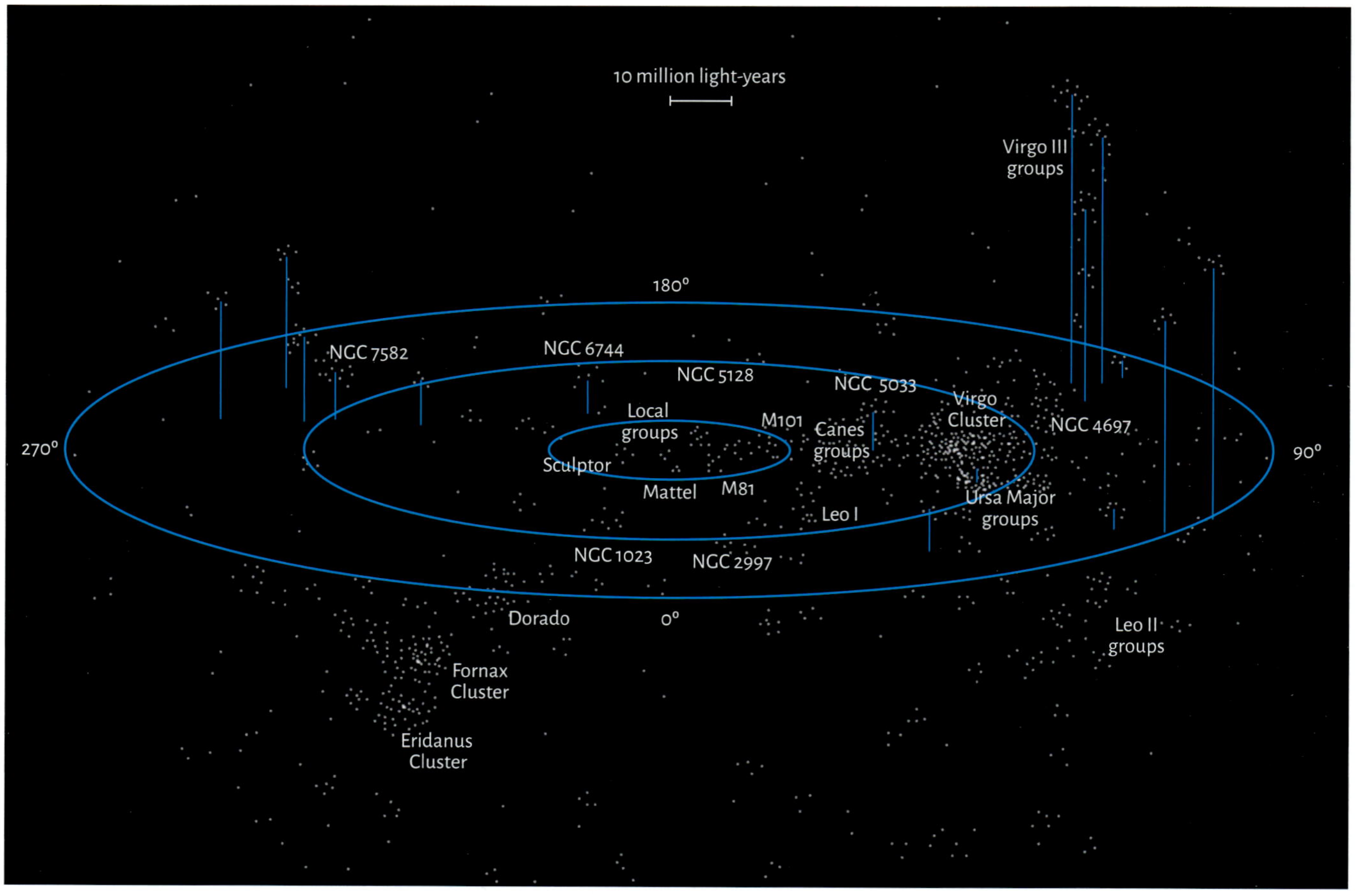

A finder chart for galaxy clusters in the local Universe.

The Fornax Cluster

The Fornax Cluster of galaxies lies at a distance of about 60 million light-years. It stretches for about 6 degrees across the night sky in the southern hemisphere (about 12 times the diameter of the full Moon) and contains around 200 galaxies. It is unusual in that it has two centres, as can be seen in the wide view of the whole cluster (next page). The galaxy in the upper left is NGC 1399, while just below it is NGC 1404. The barred spiral galaxy in the centre of the image is NGC 1365. The galaxy at lower right in the white box is NGC 1316.

We can zoom in on the upper left region, around NGC 1399. The barred spiral galaxy NGC 1365 is to the lower right. This region is generally referred to as the main centre, and NGC 1399 is the central dominant (cD) galaxy of the cluster.

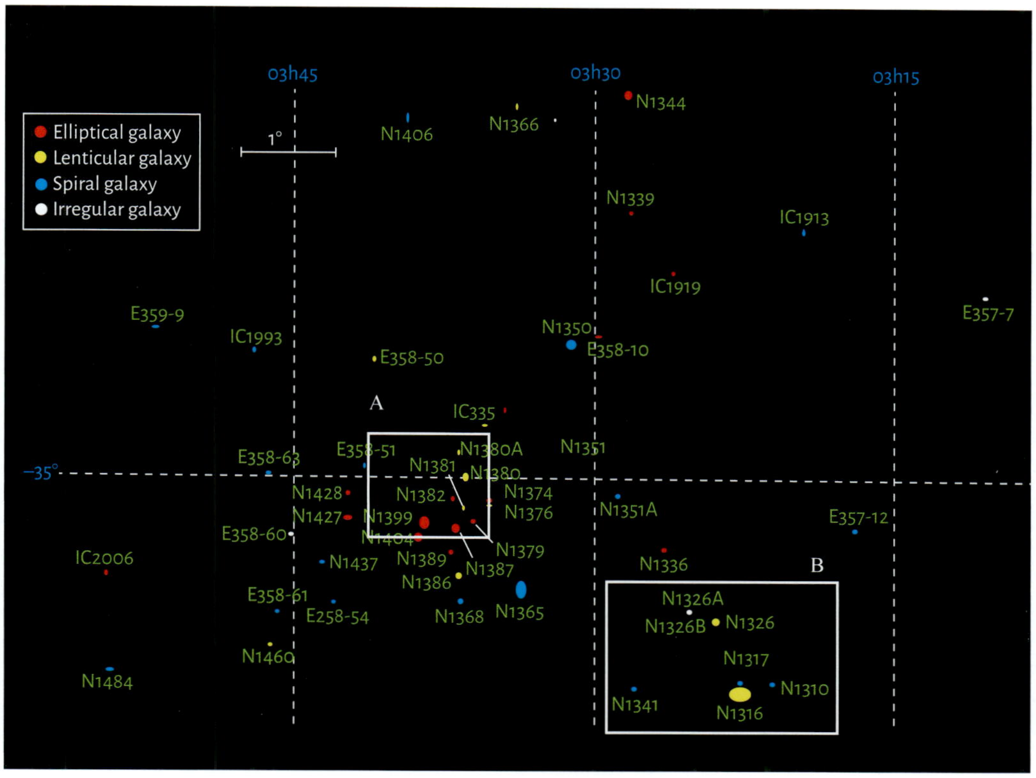

A wide-view map of the Fornax Cluster of galaxies shows that it has two centres.

In the close-up of the vicinity of NGC 1399 (below), it is clearly seen to be the dominant elliptical galaxy in the group. Numerous background galaxies can be seen through its halo. It is classed as an E1 galaxy. It has a diameter of about 120,000 light-years. It is unusual in that it has more than 5,000 globular clusters of stars around it. Most elliptical galaxies only have about 1,000, and most spiral galaxies have a maximum of a few hundred. Associated with NGC 1399 is NGC 1404, which is only about half the diameter. It is also an elliptical galaxy, this time of type E3.

The NGC 1399 region – Box A from the map of the Fornax Cluster above – taken with a red filter.

The barred spiral galaxy NGC 1365, situated near the centre of the image below, is an SBb galaxy that we discussed in Chapter 4. We mention it again here to put it in its context within the Fornax Cluster. It lies between the two centres. It is sometimes known as the Great Barred Spiral.

A close-up of the Great Barred Spiral galaxy, NGC 1365, lying between the two centres of the Fornax Cluster.

We now concentrate on the region in the lower right of the image of the Fornax Cluster, which contains NGC 1316. Here we find another rich assortment of galaxies with a wide variety of galaxy types.

NGC 1316 is classified as an SO lenticular galaxy, or sometimes as E/SO to indicate that it lies on the border between elliptical and spiral galaxies. It is known as Fornax A by radio astronomers because it is the brightest radio source in Fornax, indicating that it is an active galaxy. The dark lanes of dust are usually taken as evidence of previous mergers with spiral galaxies.

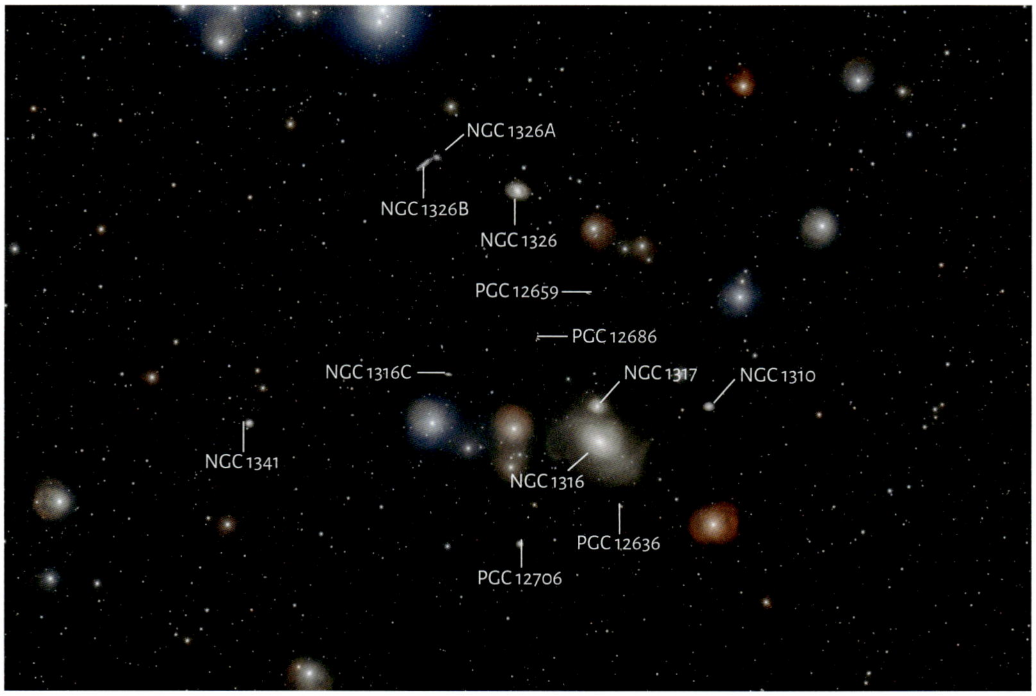

Box (B) of the Fornax Cluster image (top of previous page): a close-up of the bottom right of the cluster, containing NGC 1316.

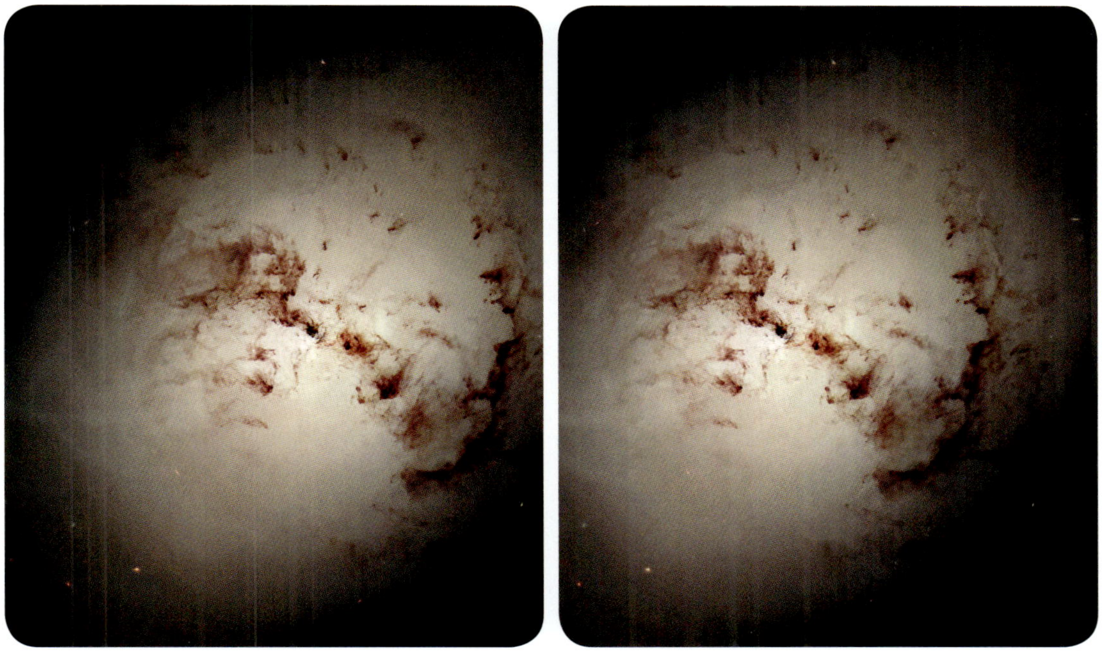

A close-up of the galaxy NGC 1316. It is classified as E/S0, but it also has unusual dark lanes across the face of the galaxy.

Another lenticular galaxy in the NGC 1316 region of the Fornax Cluster is NGC 1326. It is also classified as an S0 galaxy, but looks rather different, as can be seen here. It has a disc with no spiral arms, but it seems to have at least two concentric rings around its nucleus. S0 galaxies are normally featureless, so it is possible that the rings are remnants of a previous merger or close interaction with another galaxy.

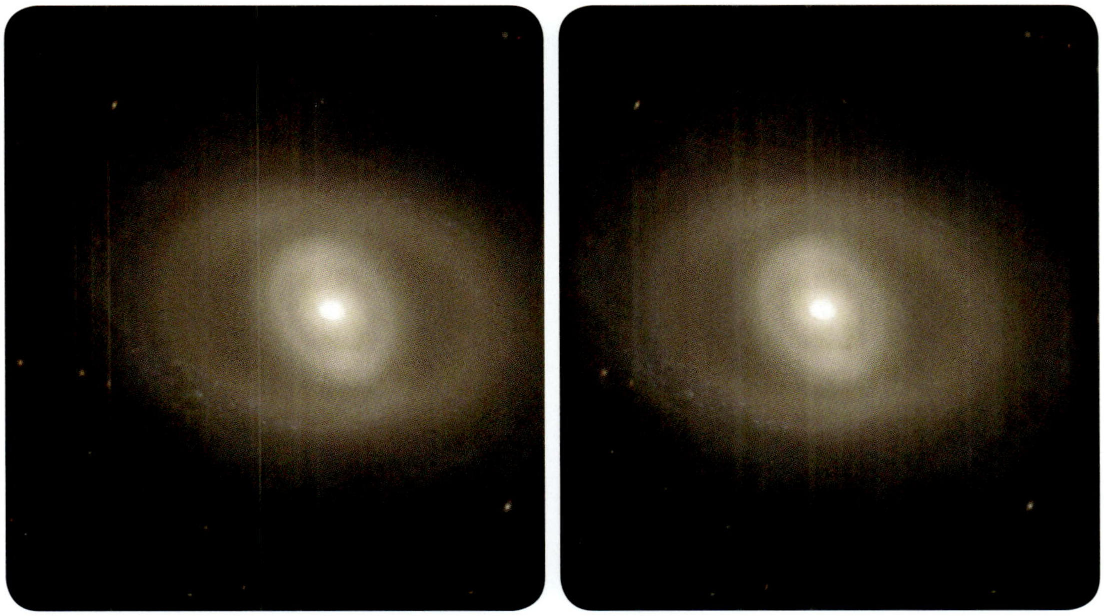

A close-up of the galaxy NGC 1326. It is classified as S0, but it also has rings that may be the result of previous interactions with one or more other galaxies.

Abell 370

Abell 370 is a far more distant galaxy cluster, situated at some 4–5 billion light-years away in the constellation Cetus. It was noted by George O. Abell, who made a catalogue of distant galaxy clusters. Abell 370 is about the most distant galaxy cluster in the Abell Catalogue, so it makes an interesting counterpoint to the Fornax Cluster.

George O. Abell, who created the catalogue that bears his name, was educated at Caltech and taught at the University of California, Los Angeles. He lived his entire life in Los Angeles, working at the Griffith Observatory and the Mount Palomar Observatory in San Diego. He used the Palomar Observatory Sky Survey to prepare the first version of his catalogue of the northern sky.

Abell Catalogue

The Abell Catalogue of Rich Clusters of Galaxies was created by George O. Abell (1927–83), and the final version was published posthumously in 1989. An earlier edition, covering only the northern sky, was published much earlier in 1958.

Abell 370 is a very rich cluster of several hundred elliptical and spiral galaxies, mostly similar to the galaxies we see around us in the nearby Universe. But it also has interesting arc-like shapes visible in the image. These are due to gravitational lensing (see Chapter 2).

In this case the bright arcs are images of even more distant galaxies whose light has been bent and amplified by the immensely strong gravitational field around this massive cluster of galaxies.

The Abell 370 Galaxy Cluster (above and next page) comprises hundreds of spiral and elliptical galaxies. This Hubble Space Telescope image exhibits arc-like shapes characteristic of gravitational lensing.

SUPERCLUSTERS

Superclusters are among the largest known features in the known Universe. They consist of assemblies of clusters and groups of galaxies into even larger entities. As already noted, our local supercluster is the Virgo Supercluster. Superclusters were first postulated by George Abell when he was making his catalogue.

The Virgo Supercluster

The Virgo Supercluster is centred on the Virgo Cluster and includes our own Local Group and groups on the other side of the Virgo Cluster known as Leo II and Virgo III; it extends all the way to the Eridanus and Fornax clusters. The gravitational centre is undoubtedly the Virgo Cluster, and all the other components of the supercluster are moving under the influence of this cluster's gravity.

The definition of superclusters is that they must be self-gravitating, that is, under the gravitational influence of the central region. This is quite staggering when it is realized that a supercluster can be well over 100 million light-years across.

We know that gravity is a long-range force that cannot be blocked. Unlike electrical charge, gravity is always positive - there is no such thing as negative mass. Electrical charge can be positive or negative, and magnetic poles can be both north and south. Gravity is always additive. Consequently, if there is enough mass, gravity will dominate galactic structure over large distances. The Virgo Supercluster contains almost 50,000 galaxies, so there is certainly enough mass!

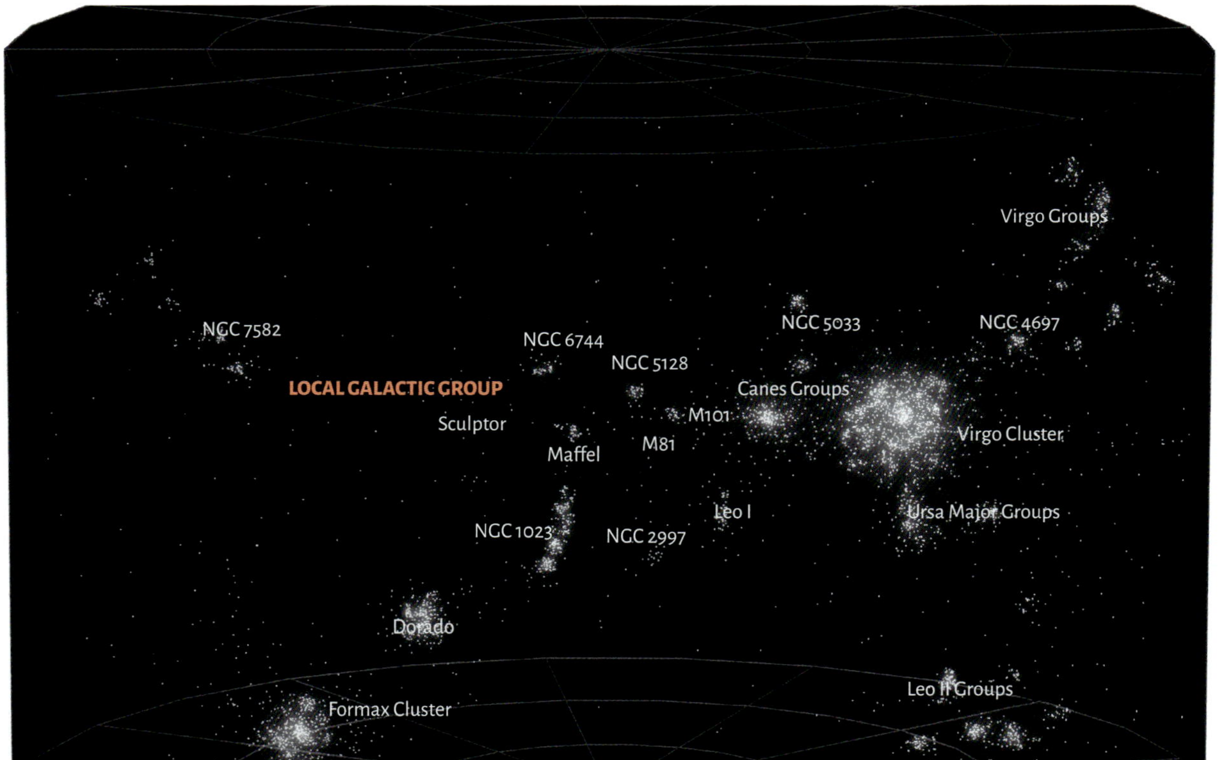

The Virgo Supercluster of galaxies.

Laniakea and the Great Attractor

There is an even larger structure of which the Virgo Supercluster is a part. In 2014 a group of astronomers, led from the University of Hawaii, published a paper in *Nature* magazine claiming that the Virgo Supercluster is part of a larger structure that they named Laniakea. This is a Hawaiian word meaning 'open sky' or 'immense heaven', chosen in honour of traditional Hawaiian navigators who travelled across the Pacific Ocean using the stars to guide them.

The team used a new technique to analyse the positions and velocities of thousands of galaxies in immense detail, to study their motions relative to each other, and discovered that there was a coherent structure containing over 100,000 galaxies, stretching some 250 million light-years across the Universe, that includes our own Milky Way Galaxy.

Laniakea encompasses the previously known superclusters Virgo, Hydra, Centaurus and Pavo-Indus as well as the clusters Fornax, Eridanus, Antlia and Pegasus, among many others. It appears to be centred on an object that is known as the Great Attractor, first hypothesized by a group from Cambridge University, UK, led by Donald Lynden-Bell, a famous 20th-century British astronomer who published a paper about this in 1988.

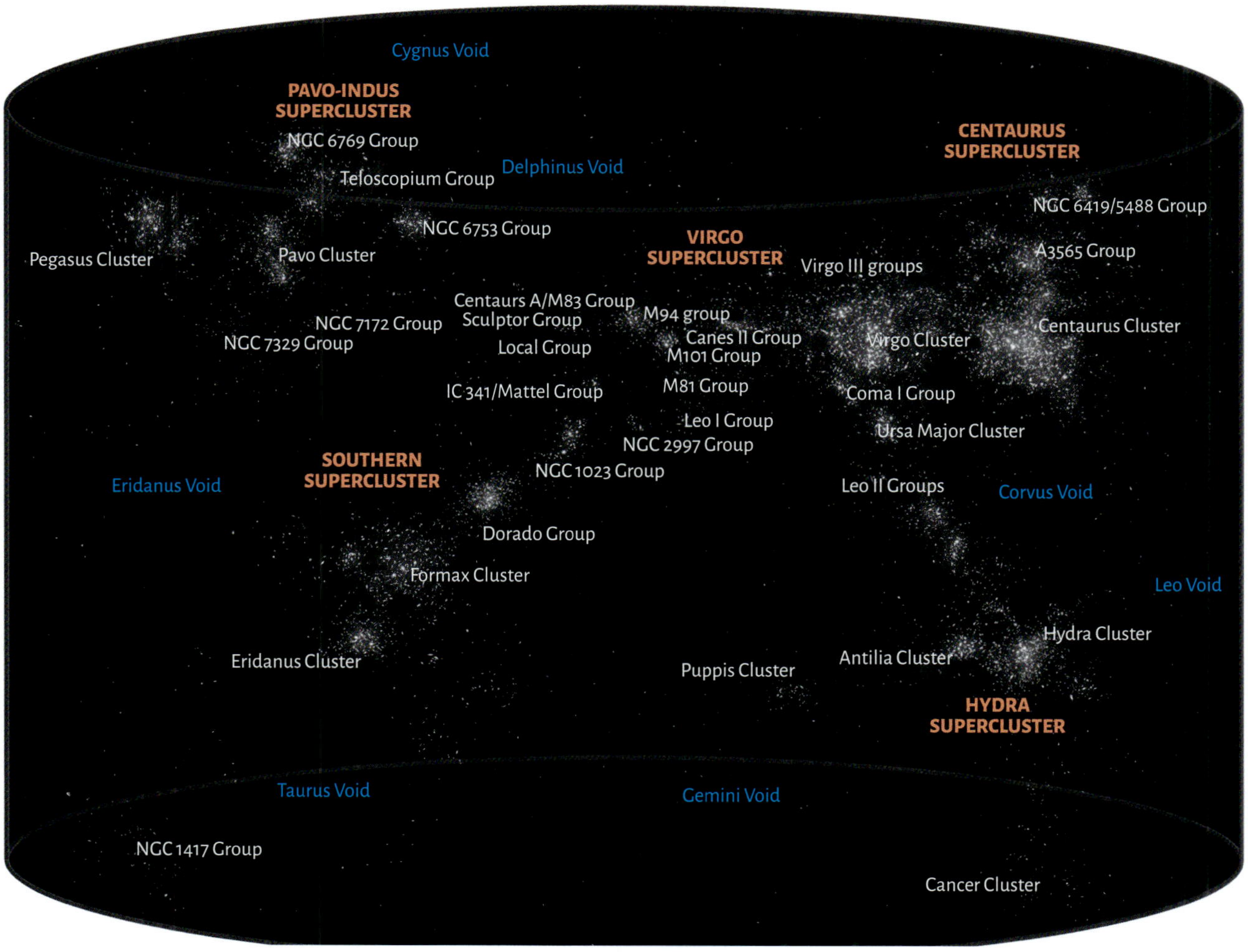

Sky map of the Laniakea structure of superclusters.

DONALD LYNDEN-BELL (1935–2018)

Donald Lynden-Bell was a 20th-century British astronomer, who was President of the Royal Astronomical Society from 1985 to 1987. He was part of the team who first proposed that the Milky Way Galaxy must have collapsed from a single giant gas cloud. He won many awards for his other theories and discoveries, including variables that came to be known as T Tauri stars. He led the paper that first suggested the existence of the Great Attractor.

Donald Lynden-Bell (1935–2018).

The Great Attractor was estimated to be 50 million billion (a five with 16 zeroes after it) times the mass of the Sun, which is equal to the mass of a typical supercluster. We cannot see this object, since it is obscured from us behind the bulk of our Milky Way Galaxy. However, we know that it is there, based on the motions of the galaxies under the influence of its gravity.

As we have noted, it is very difficult to see through our own Galaxy due to the very many stars and, more importantly, all the dust that lies in the plane of the Milky Way that effectively obscures everything behind it. The map of galaxies across the sky (below) contains a large empty band running from top to bottom that is devoid of background galaxies because they are hidden from us by dust.

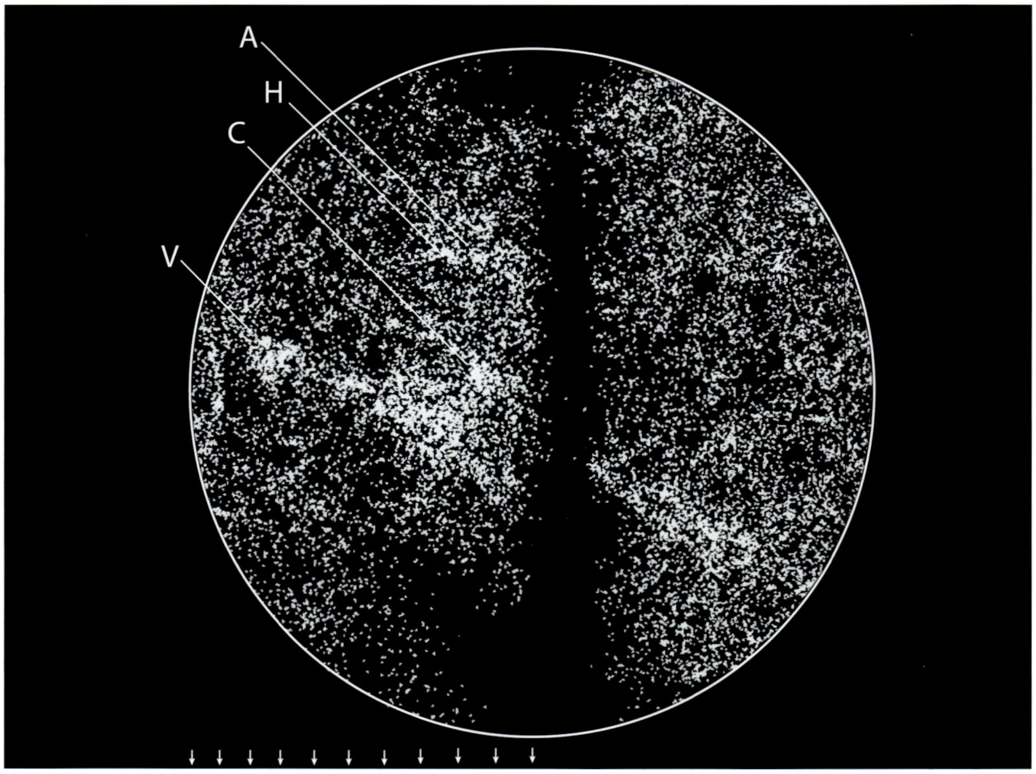

Map of galaxies taken from the original paper showing a hemisphere of the sky centred on the Great Attractor. Every dot is the plotted position of a galaxy. The Virgo, Centaurus, Hydra and Antlia clusters are all indicated by their first letters.

This empty band is in the plane of the Milky Way Galaxy, and is known as the 'Zone of Avoidance' because galaxies appear to avoid it, but in reality we just can't see them.

It is frustrating that the Great Attractor lies there, because it means we cannot see it directly, but every galaxy, including the Milky Way, can feel its gravitational influence. The Great Attractor is believed to be the central point of the Laniakea Supercluster of galaxies.

The Perseus-Pisces Supercluster

The next nearest supercluster after Virgo is known as the Perseus-Pisces Supercluster. As one might expect, when viewed from our home planet it straddles those two constellations. It is roughly 250 million light-years away and extends for over 40 degrees across the sky, meaning it is over 150 million light-years in extent. Furthermore, some astronomers include additional objects in the supercluster and claim it extends to nearly 300 million light-years.

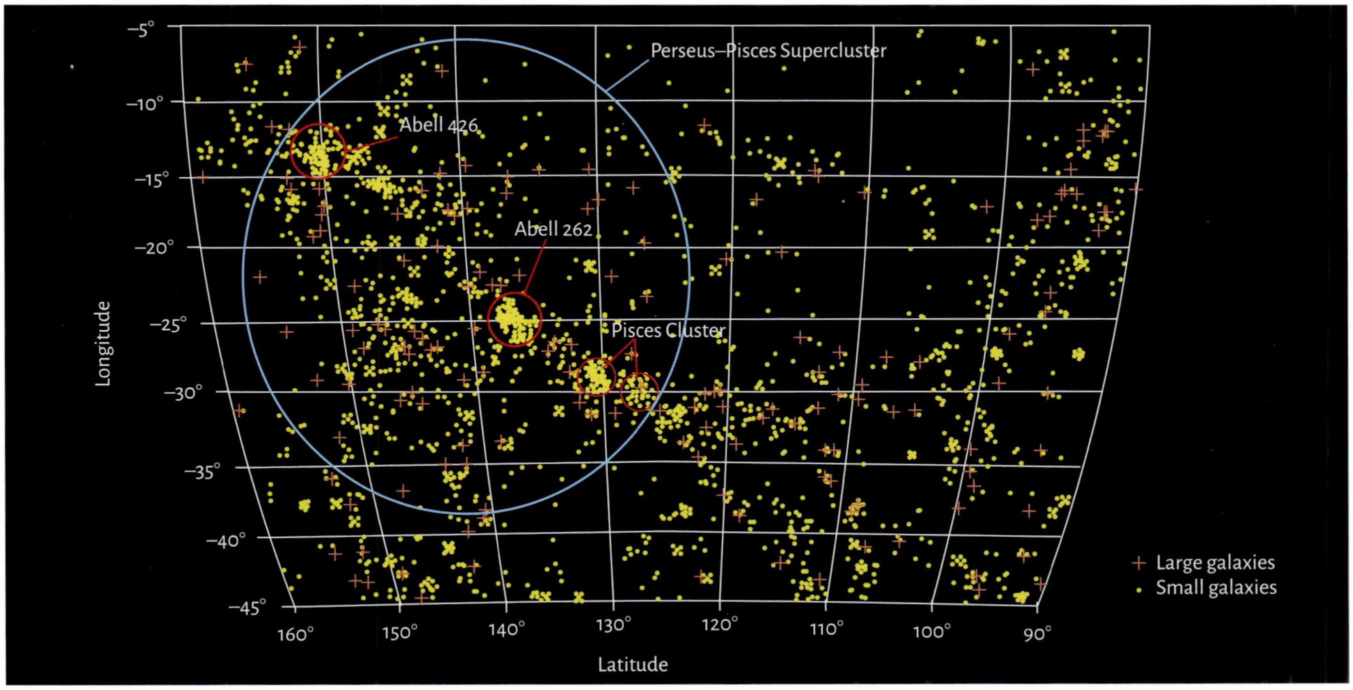

Map of the Perseus–Pisces Supercluster.

The Perseus Cluster, Abell 426, at one end of the Perseus-Pisces Supercluster.

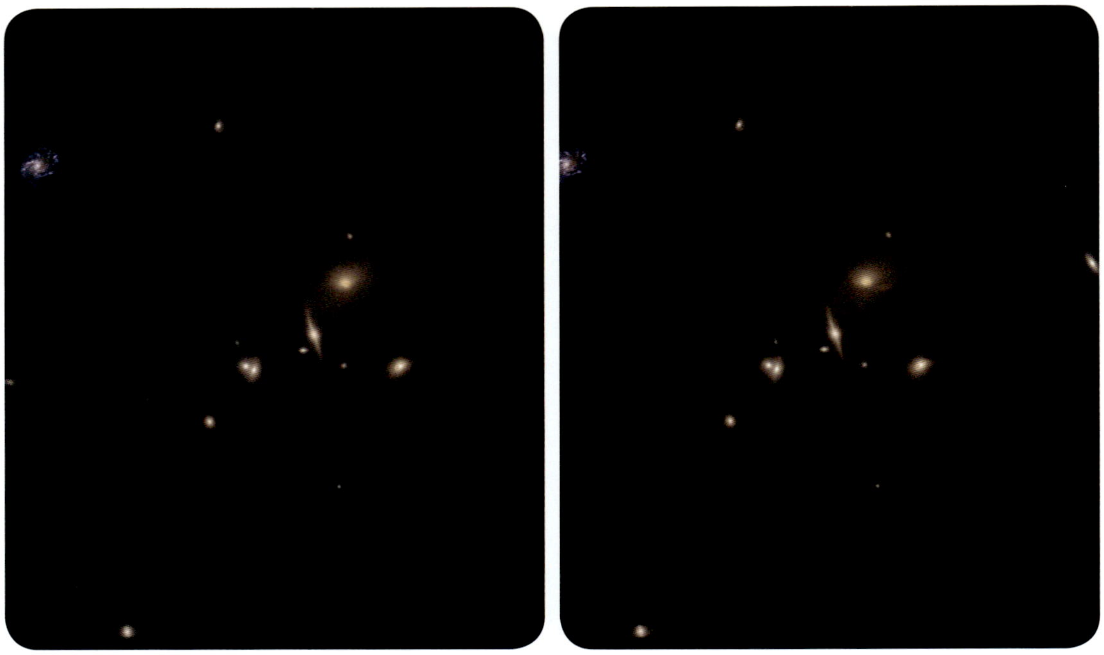

The Abell 262 cluster, near the middle of the Perseus-Pisces Supercluster.

In the sky map of Perseus-Pisces (previous page), the blue circle indicates the approximate extent of the supercluster. The crosses, dots and plus signs represent galaxies and clusters of different sizes. The red circles represent well-known clusters, with the Perseus Cluster (Abell 426) at top left and the two components of the Pisces Cluster at lower right. The red circle nearer the centre is the cluster Abell 262.

The Perseus Cluster contains thousands of galaxies, while Abell 262 contains the giant elliptical galaxy NGC 708 as the central dominant (cD) galaxy, but is an unusual cluster in that it contains mostly spiral galaxies. The elliptical galaxy NGC 708 can be seen left of centre in this image, just above and to the right of the edge-on spiral galaxy NGC 705.

Walls, Voids and the Cosmic Web

Returning briefly to the map of the Perseus-Pisces Supercluster, it is striking that a large fraction of its components lie along a linear feature that stretches from upper left to lower right. Above and to the right of this feature there is a distinct lack of galaxies, as there is below it.

Redshift Surveys

The amount by which the light from a distant galaxy is moved towards the red end of the spectrum is dependent on the speed with which it is moving away from us. This is the famous 'redshift' (or 'Doppler shift') that we met in Chapter 2. Astronomer Edwin Hubble showed that the amount by which the light from a distant galaxy was redshifted was proportional to its distance away from us. The constant of proportionality is known as the Hubble constant. Hence, redshift surveys are where astronomers measure large numbers of redshifts of distant galaxies to determine their distances, and thereby make a three-dimensional map of the Universe.

Linear features turn out to be a common aspect of structures on the scales of superclusters. Astronomers refer to these features as 'walls' or 'filaments'. We note that since a filament is essentially the shape of a drinking straw, it is very difficult to tell the difference between a filament and a wall that is seen edge on.

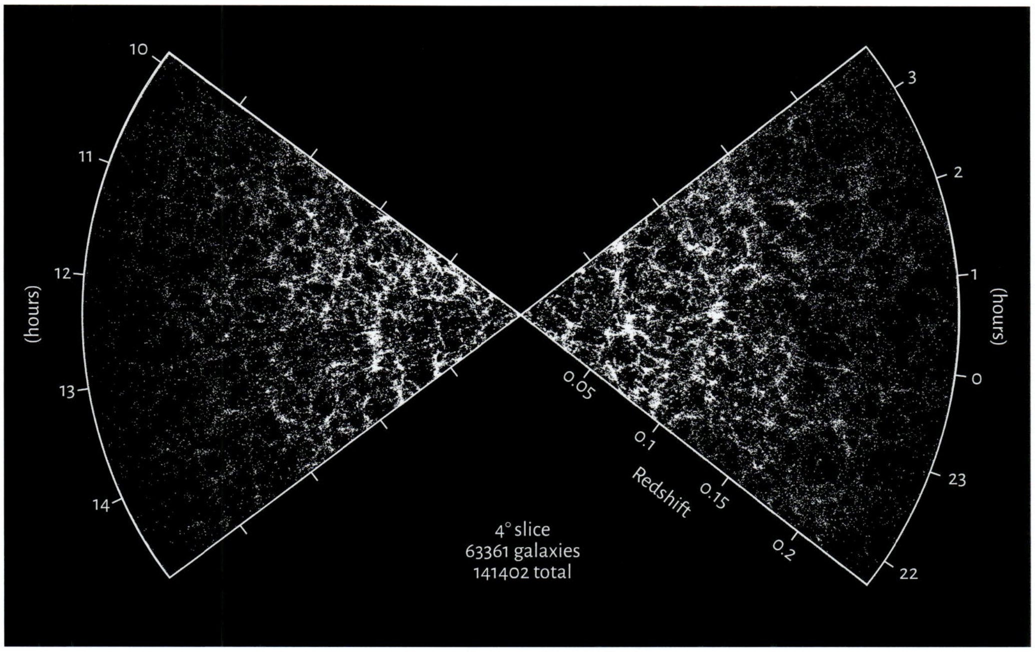

Slices through the Universe, drawn using the Two-Degree Field Galaxy Redshift Survey (2dF). The plot extends to a maximum distance of about 3 billion light-years and illustrates the cosmic web of voids and filaments – structures over 100 million light-years in extent.

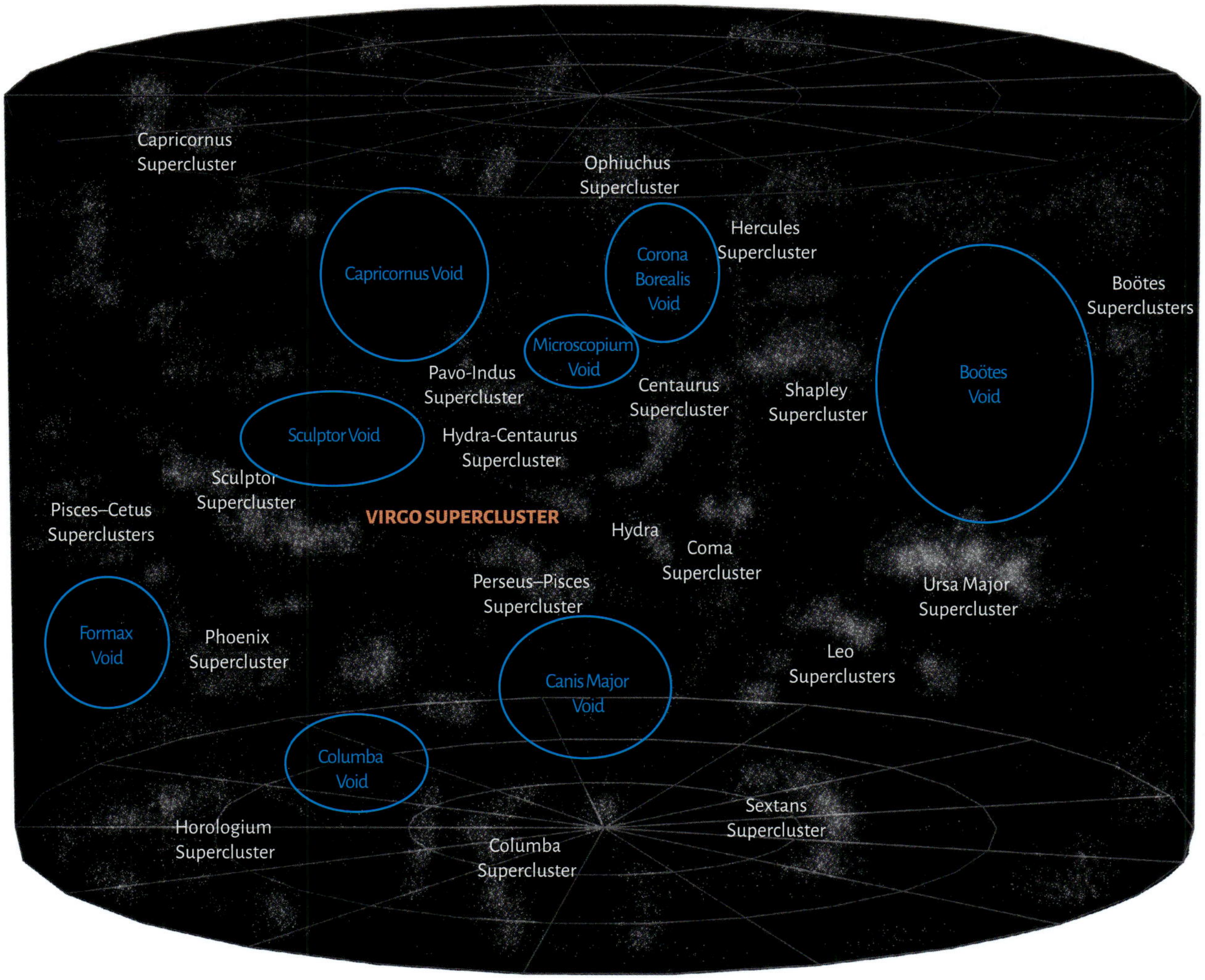

Our local part of the Universe, showing superclusters and voids.

It further transpires that between the Virgo Supercluster and the Perseus-Pisces Supercluster there is a lack of galaxy clusters. This paucity of galaxies is known as a 'void', and in this case it is known as the Taurus void. These walls, filaments and voids are common features of the largest-scale structures in the Universe, which we now explore.

In 2001 an international team, led by Edinburgh-based astronomer John Peacock, published a paper in *Nature* magazine containing the results of a redshift survey that they had carried out using the Anglo-Australian Telescope at the Siding Spring Observatory, Australia, of more than 140,000 galaxies. They measured every one of these galaxies' redshifts. The survey had a 2-degree field of view, so it was known as the 2dF.

In the map from the Peacock paper, which is reproduced at the top of the previous page, we on planet Earth sit at the centre point where the two triangles meet. The triangles are slices through a cone of the sky in each of two opposite directions. The angles marked around the outside edge represent positions in the sky,

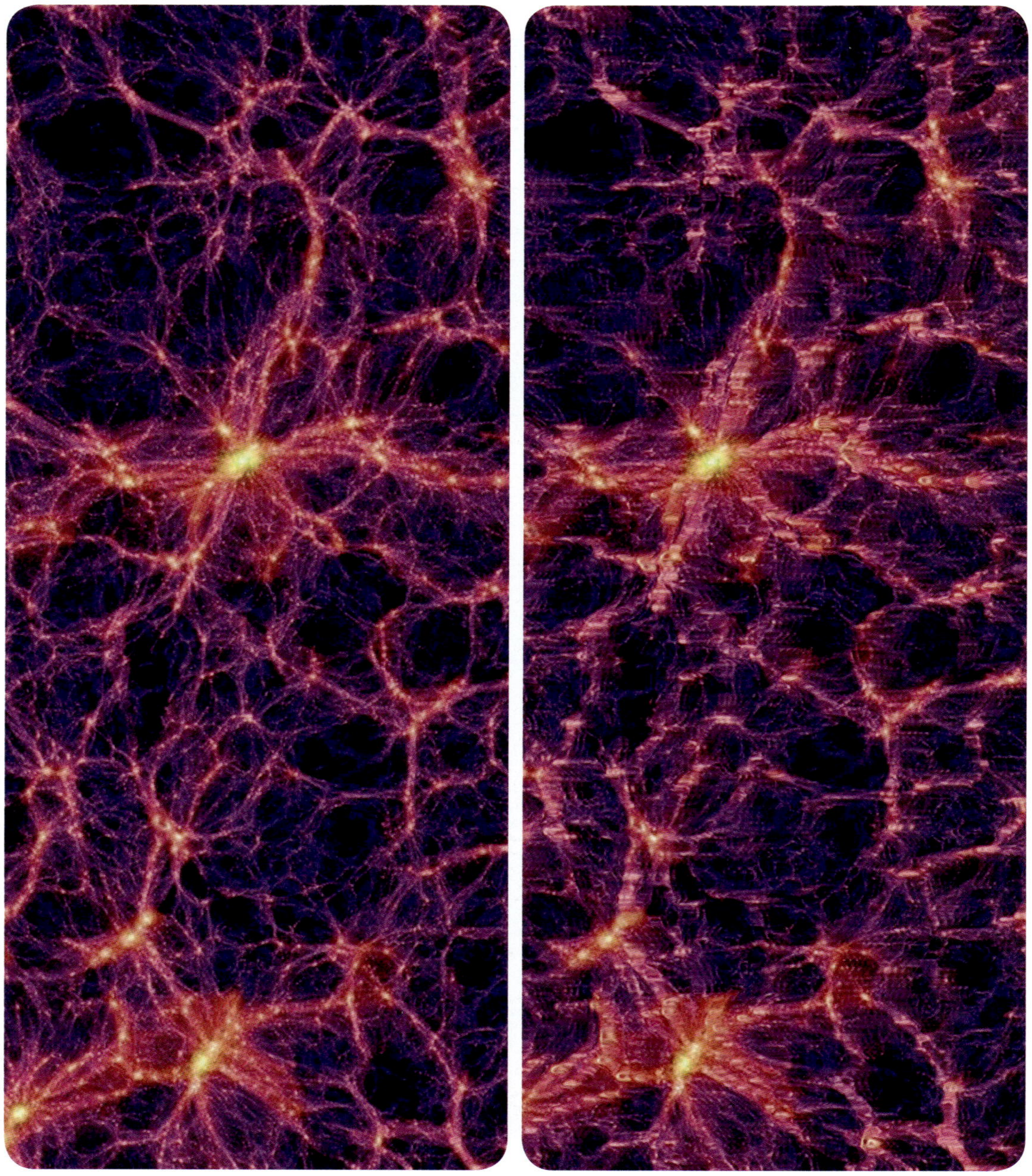

The cosmic web simulated by a supercomputer.

and the radii represent redshift. Every point plotted in this map is a measurement of a galaxy's position in the sky as an angle (the direction in which we see the galaxy) and its redshift, which is plotted radially as a proxy for the distance away from us. To put the redshift scale in context, a redshift of 0.15 in radius on this plot corresponds to a distance from us of about 1.5 billion light-years.

What stands out very clearly from the 2dF Galaxy Redshift Survey is that the galaxies are not evenly spread out. They cluster, and they extend in the long filaments or walls that we have noted. In addition there are very large voids between the filaments, which are rather like bubbles in space, exactly as seen in the computer simulation on the previous page.

In the diagram of the relatively local Universe (two pages back, lower image), we see the local clusters, superclusters, walls, filaments and voids labelled as astronomers name them. We are in the Virgo Supercluster, indicated in the middle. Note the Centaurus and Pavo-Indus superclusters that are next to the Virgo Supercluster in this diagram. Compare this with the positions of those superclusters in the earlier diagram of Laniakea. This shows the vast difference in scale between this map and our previous diagrams.

Astronomers use the term 'cosmic web' to describe what is seen of filaments and voids because they form something like a three-dimensional spider's web. They find that this is consistent with the predictions of a model in which the three-dimensional distribution of matter across the Universe collapses under gravity to form voids, walls, filaments, superclusters, clusters, galaxies, stars and planets.

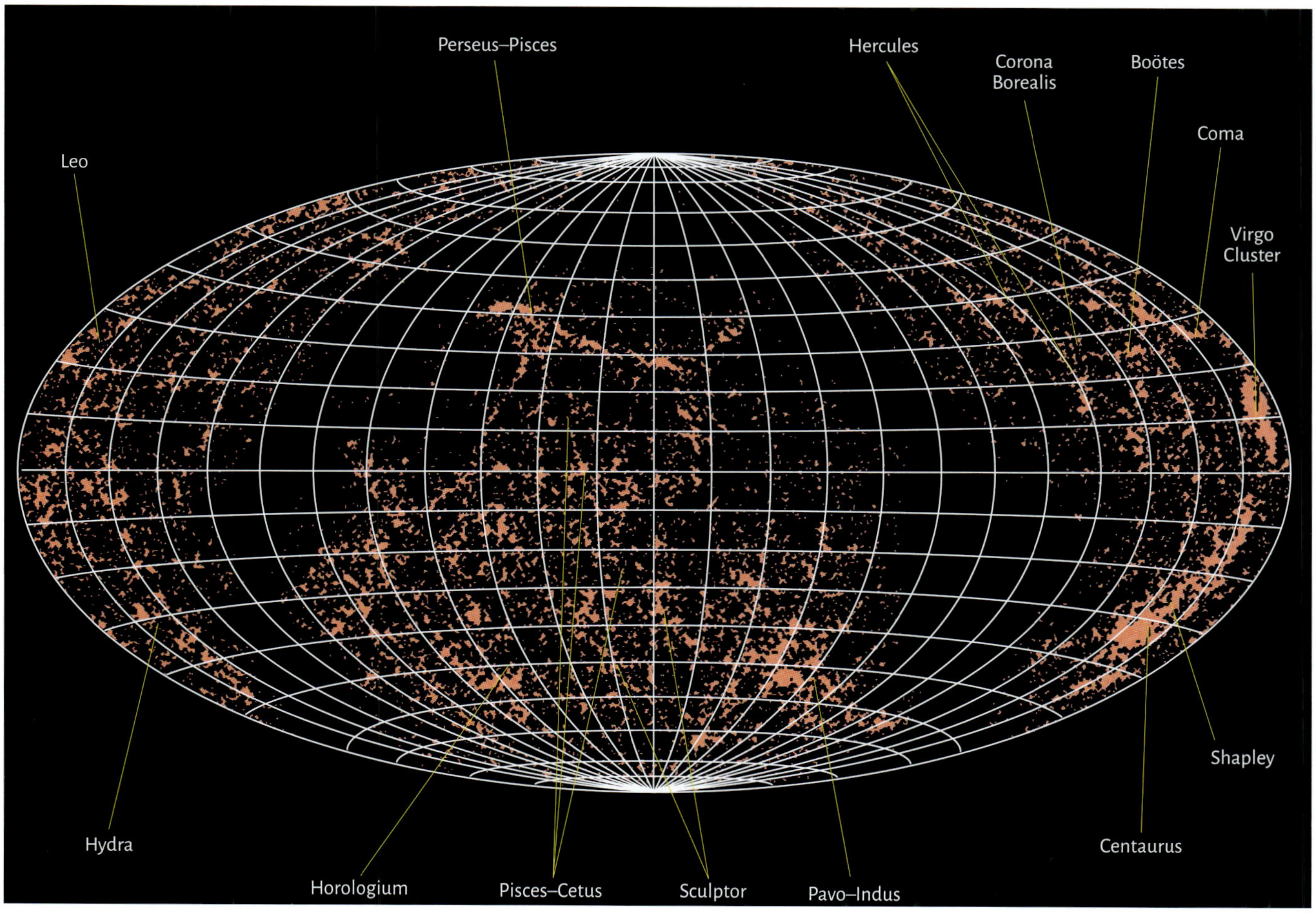

The Universe within 1 billion light-years of us. The positions of various clusters are marked.

In the computer models, the filaments channel matter into superclusters and clusters that form at the intersections (or nodes) where two or more filaments meet. However, these models of matter forming filaments and then flowing along filaments onto nodes only work if astronomers include dark matter (see discussion in Chapter 2) in the models.

We show the results of one of the models two pages back. This is the output from a supercomputer, which was given a set of constraints covering the laws of physics, including those relating to dark matter, and which was programmed to calculate how matter would move under gravity. The simulated large-scale Universe looks remarkably like what was seen by the 2dF Galaxy Redshift Survey. The galaxies have clustered, and some of the clusters have themselves clustered into superclusters. On the largest scales, the matter has formed into filaments, walls and voids.

The image on the previous page is a map of the nearest billion light-years of the Universe and plots all the known galaxies, clusters and superclusters, marking the positions of the clusters and superclusters that we have come across throughout this chapter. It looks remarkably like the simulation.

As we pointed out in Chapter 2, the cosmological principle states that, above a certain scale, the Universe looks the same everywhere. We are starting to see this in the patterns of the voids and filaments revealed in the 2dF Galaxy Redshift Survey and in computer simulations. The astute reader will point out that the Universe does not look *exactly* the same everywhere, because all clusters are subtly different. So one analogy for the cosmological principle might be to think of the Universe as being like vegetable soup, which contains both thicker parts and thinner parts, and if examined closely enough, the differences can be seen. However, by and large it is difficult to tell one bowl of vegetable soup from any other. This is how the cosmological principle views the Universe: like a giant bowl of vegetable soup which, by and large, is the same wherever we look.

The Largest Structures in the Universe

This then leaves us with the question of whether there is any larger structure to be discovered in the Universe. Can we find structures that span a significant fraction of the observable Universe?

In trying to answer this question, astronomy doctoral student Alexia Lopez announced in 2024 that she and her supervisor, astronomer Roger Clowes, both from the University of Lancashire, UK, had discovered one of the largest structures in the Universe. They called this the Big Ring. They had previously announced the discovery of another large structure known as the Giant Arc. The two structures are shown in the pictures above, and their relative positioning is illustrated in the diagram on the previous page.

The ring and the arc are structures of galaxies at huge distances from us. Both images (below and top of next page) are about 3 billion light-years across, and the objects are seen at a distance of about 9 billion light-years. The Big Ring is over a billion light-years in diameter. These distances are significant fractions of the size of the entire observable Universe.

This represents a significant challenge to the cosmological principle – and the analogy we give above of the Universe being like vegetable soup. Structures this large should be rare, if seen at all. They are larger than the estimated scale on which the Universe should appear homogeneous.

Both the Big Ring and the Giant Arc lie in the same part of the sky. Again, there is no explanation for this. Consequently, astronomers are currently debating what these structures are and how they came into existence. As yet there are no answers.

Alexia Lopez, a postgraduate student from the University of Lancashire, UK, who discovered the Big Ring and the Giant Arc.

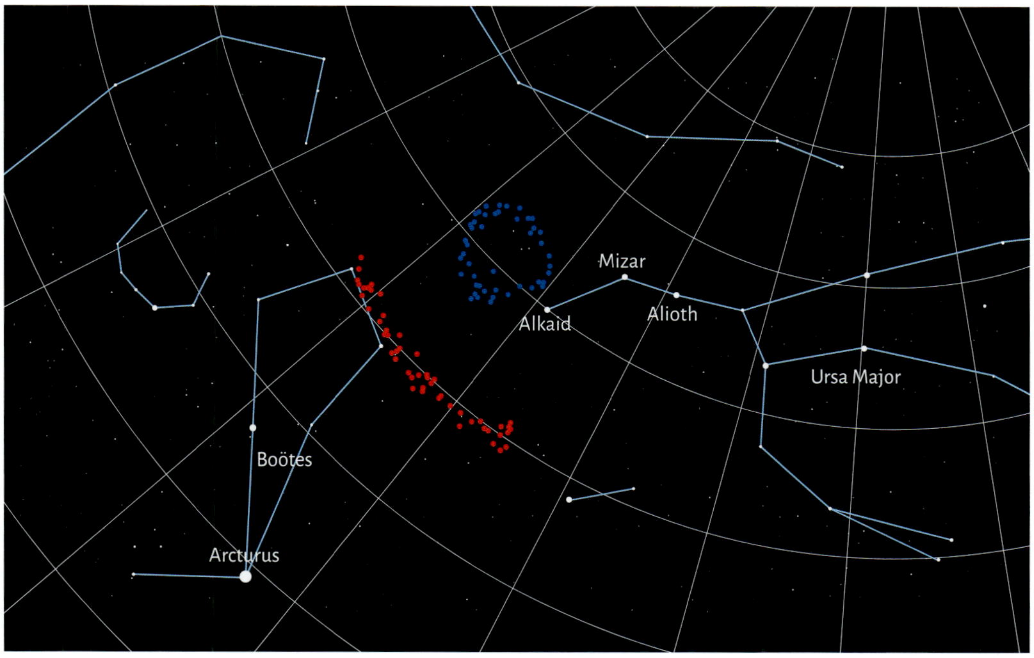

Sky map showing the relative positions of the Big Ring (blue dots) and Giant Arc (red dots). The usual foreground constellations are shown for orientation.

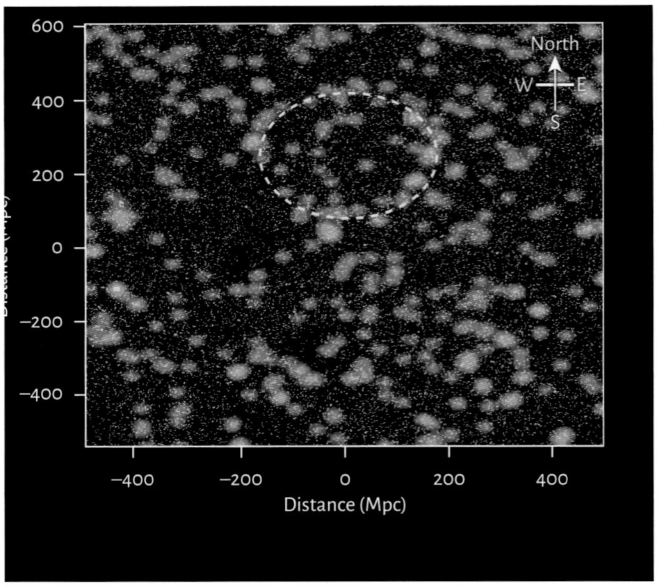

The Big Ring is the circle of grey dots just above the centre of this image, forming a ring shape.

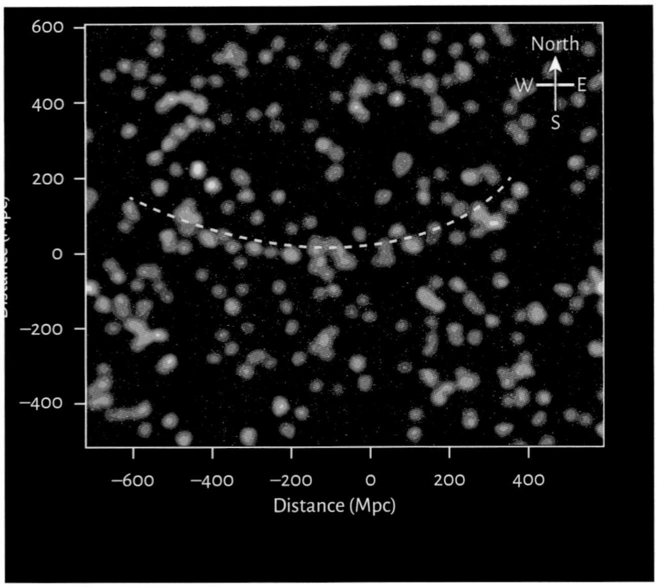

The Giant Arc is the curved arc of grey dots that can be seen across the middle of this image.

How the structures were discovered

The Big Ring and the Giant Arc were discovered by analysing the spectra of light from distant quasars. Quasars are used because between them and us there are many ordinary galaxies that we cannot see by their own light because they are too faint, but they can absorb the light from the background quasars. When an object contains a particular element and a light is shone through it, that element absorbs light at a series of very particular signature wavelengths, which can be measured in a laboratory. At a great distance, the wavelength of light that it absorbs is red-shifted. So it is possible to select objects at a very specific distance by looking for the absorption at a particular red-shifted wavelength. The images show the position of the gas appearing in a thin slice of space 9 billion light-years away. This makes the Big Ring and the Giant Arc genuine structures at a specific distance in space.

Quasars

Quasars are among the brightest objects in the Universe, so they can be detected at vast distances. They are a subset of what we know as active galaxies: they have a black hole at their centre, powering the most impressive jets of extremely high luminosity that can be seen right across the Universe.

So we have finally reached the furthest objects in the part of the Universe we can see, and have reached the limits of our understanding in modern astronomy. We hope you have enjoyed the journey.

Islands in Infinity

This image was captured by the James Webb Space Telescope using over 12.5 hours of exposure time. It contains an estimated 25,000 galaxies – or 'islands in infinity', as we have named them – seen in a distant cluster of galaxies some 4.6 billion light-years away. But the field of view it covers is only the size of a grain of sand held at arm's length, so about 26 million such images would be needed to cover the entire sky. The arcs in the image are background galaxies affected by gravitational lensing, as discussed earlier in this chapter.

How I Made the Stereos

To perceive an object in three dimensions, our two eyes must view it from slightly different angles. When the distance from us to the object becomes too great, it is no longer possible to form a stereo image. Distant celestial objects, such as nebulae and galaxies, are far too remote for us to see from sufficiently separate viewpoints to form a true 3-D view of them.

I have been photographing the night sky from my private observatory for around 30 years. In 2005, I began to wonder how distant celestial objects might appear if we could see them three-dimensionally. When people view beautiful images of nebulae, star clusters or galaxies, they often appear like flat paintings on a canvas. It is easy to forget that these are three-dimensional structures suspended in space.

For as long as I have captured images of celestial objects, I have perceived them as three-dimensional in my mind. Scientific data makes these internal visions far more accurate, and the 3-D technique I have developed allows me to share those visions with others.

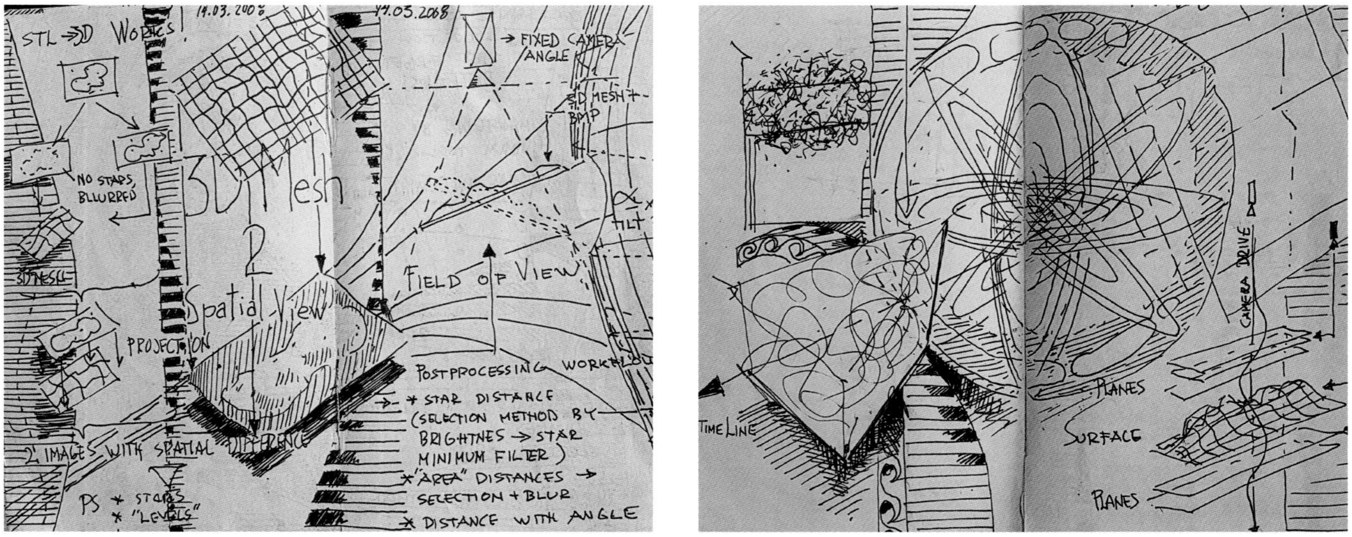

Pages from my small Moleskine notebook drawn in 2008 show early sketches and plans for the final 3-D conversion technique.

A stereo close-up of NGC 1055 reveals the complex structure of its dark dust lanes. Several distant background galaxies are also visible.

Driven largely by curiosity, I developed a method that enables me to transform any celestial object into a stereo pair, as well as into various 3-D image formats and models. Before beginning the 3-D conversion workflow, I gather as much scientific data as possible. In addition to distance information, many other scientific clues are essential for building a realistic simulation of a 3-D structure.

Cosmic structures, such as nebulae and galaxies, form in accordance with the laws of physics. As a result, similar patterns recur throughout the Universe, each time with subtle variations. Additional small clues can often be discovered by carefully studying the image itself. For example, if a dark dust lane is visible in a galaxy, it must lie in front of the brighter regions – otherwise it would not be visible at all.

The accuracy of my 3-D interpretations depends on the amount of information available and how well I can apply it. The final 3-D image is always an informed simulation of reality, based on established scientific facts, logical deduction and a touch of artistic creativity. In essence, this is a form of sculpting, but on a cosmic scale.

Initially, I believed that galaxies were relatively simple in structure, consisting mainly of a disc and a central bulge. However, as I began the 3-D modelling work, I encountered a remarkable variety of forms, interactions and dynamic structures.

Hubble Ultra Deep Field converted to a 3-D stereo. All foreground stars have been removed; only galaxies remain in this image. For some of them, I had distance data available to use, but for most, I did not. In those cases, I used redness as a distance indicator. The farther a galaxy is from us, the faster it is receding, which results in redshift – discussed in Chapter 2.

There are violent expulsions of gas and energy, gravitational tidal forces, distorted stellar fields and gas streams, colliding systems, and much more. Translating these phenomena into three dimensions and then viewing them in stereo for the first time was an extraordinary experience.

The process of converting galaxy groups into 3-D representations was highly labour intensive, but the results were immensely satisfying. Seeing the relative distances and spatial relationships revealed the true three-dimensional choreography of these interacting galaxies, opening up an entirely new way of understanding their structure and behaviour.

Directly edge-on galaxy NGC 7814 and background galaxies.

We live inside the Milky Way. Everything we observe beyond it is seen through the dense star fields of our own Galaxy. These foreground stars can be so numerous that they obscure the celestial objects lying behind them. To overcome this, I decided to take an unorthodox approach: I carefully removed all the foreground stars of the Milky Way.

As a result, we are now able, for the first time ever, to visualize what space might look like if we could step just outside our home Galaxy - a view dominated by galaxies and galaxy groups.

To remove the stars of the Milky Way, I used statistical methods. However, this process also tended to erase many of the faint, almost point-like, background galaxies, as well as some delicate details within larger galaxies. To solve this, I had to resort to labour-intensive manual work.

First, I created two versions of the same image: one with the foreground stars intact, and one with them removed. In the second step, I created a layered composite from the two versions.

The left-hand side shows the original photo of the Abell 262 galaxy cluster. The right-hand side shows the same image with all foreground stars removed. This is how the cluster might appear if we were ever able to travel beyond our own Milky Way Galaxy.

A close-up on a small portion of the pair of images above, both with and without stars. The image on the right contains only galaxies.

The left-hand side of this image shows the original photo of Andromeda, while the right-hand side shows the galaxy with all foreground stars removed. Most of the star-like objects visible on the right are globular clusters surrounding the galaxy. Some faint background galaxies are also visible.

A close-up from a stereo pair of M31 shows how the dark dust lanes float in front of the galaxy's disc. The small companion galaxy, M32, is located near the centre of the much larger M31 and can also be seen floating in front of the galaxy's disc, towards the upper right. Note the beautiful globular clusters, seen here as small star-like objects, surrounding M31 in front of and behind the galaxy disc.

Then, zooming in on small rectangular sections of the image, I manually restored every missing background galaxy into the starless version. Typically, there were around 100–160 such areas to process manually in a single image.

After completing this meticulous step, I was left with a pristine image - one where every background galaxy had been restored and the full detail of larger galaxies preserved. The final view offers a rare and profound perspective, a glimpse into deep space, as if we had truly travelled beyond the boundaries of the Milky Way.

The Great Andromeda galaxy, M31, was one of the first objects I processed using this method. Since M31 is relatively close to us, we can see hundreds of globular clusters surrounding it. I found a research paper detailing the known globular clusters in Andromeda and used this information to isolate each one from the foreground stars - even though at this scale these clusters appear almost indistinguishable from stars.

There were also some very small and faint background galaxies within the field of view, and these are now visible in the starless version of the image.

I was surprised by how effective this method turned out to be. It clearly emphasized how close the foreground stars are compared to the immense distance of the galaxy itself. Our own Milky Way has a diameter of roughly 150,000 light-years. The stars in this image are typically located no more than a few tens of thousands of light-years away. In contrast, M31 lies at a distance of about 2.54 million light-years.

Between the foreground stars and Andromeda, there is essentially nothing visible - there are only photons racing through space and other invisible radiation, along with a little gas and dust. This also means that if we were somehow able to travel just outside our Galaxy, the view of Andromeda would remain almost exactly the same.

The Milky Way

Every year, fewer people are able to see the Milky Way - the soft band of dim light stretching across the night sky - because of increasing light pollution. When we look at the Milky Way, we are seeing the inner edge of the disc of our home Galaxy.

In 2021, I completed a massive imaging project of the northern Milky Way. I began planning the project between 2007 and 2008, and started capturing images in 2009. The entire process took 12 years to complete. The result is the only photograph ever made that captures the northern Milky Way with such depth and detail.

One of the key motivations behind creating this monumental panorama was the simple fact that no such image existed and I had a personal need for a photograph like this. I intended to use it as a map for future adventures.

The final image required a total of 1,500 hours of exposure time and is composed of 301 individual frames. It covers an area of 145 by 22 degrees of sky. The original resolution of the panorama exceeds 120,000 by 18,000 pixels.

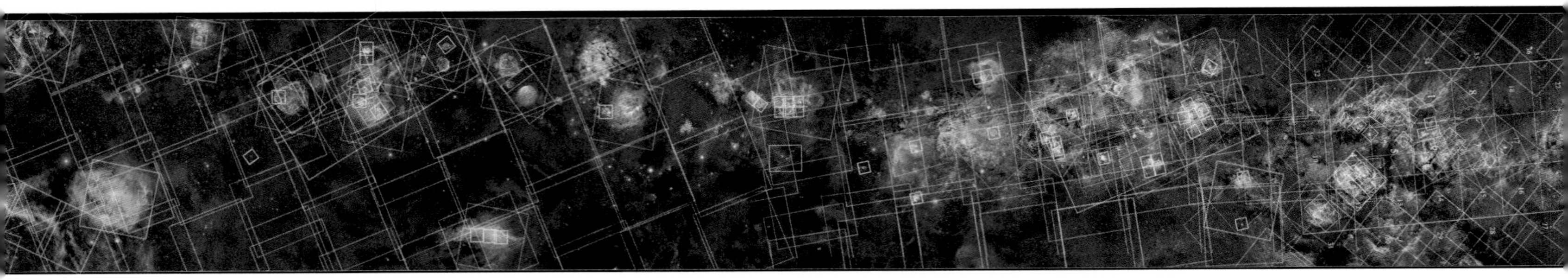

Every frame used for the mosaic is marked in this image.

Because each colour channel was captured separately at full resolution, the panorama is as accurate as a standard digital photograph would be if it had a resolution of 240,000 by 36,000 pixels.

Beyond the exposure time, countless hours were spent assembling the image, with every mosaic panel seamlessly stitched together. Many of the sub-images and mosaics are also artworks in their own right, resulting in a highly complex and multilayered mosaic structure.

The photographed area captures a vast section of the northern Milky Way in high resolution. Beyond the image's immense scale, it is remarkably deep, revealing extremely faint and previously unphotographed nebulae scattered across the galactic plane.

To manage a project of this magnitude, I had to develop many new working methods, and they all had to be in place before I could begin. Once the imaging started, the overall plan could not be changed without risking the entire project. Everything needed to be precise, from the composition and its alignment with the Milky Way's structure, to numerous technical aspects, such as handling data from different optical systems with varying spatial resolutions. I will not delve too deeply into the technical details, as the complex techniques involved are simply tools that enable me to create my art.

Astronomical photography is an incredibly time-consuming process. To produce a colour image, I must photograph each target at least three times - once for each colour channel - using different filters. Exposure times can be extremely long: for some of the faintest targets, I have spent more than a hundred hours on a single image. On average, each panel in the mosaic required around 25 hours of exposure. Every detail had to be carefully planned.

I began sketching out imaging plans over a decade before the project was completed, jotting down the first ideas in my small black Moleskine notebook in 2008. Even then, I knew the project would take more than ten years, but that didn't trouble me. I love long-term projects; they give purpose and direction to my work as an artist.

I often compare these long imaging efforts to a relationship. That is exactly how it felt when stitching the mosaic together, and some pieces did not fit as I had hoped. I had to go back and reshoot those parts, which could take months, even years. But in the end, everything came together smoothly, with no visible seams.

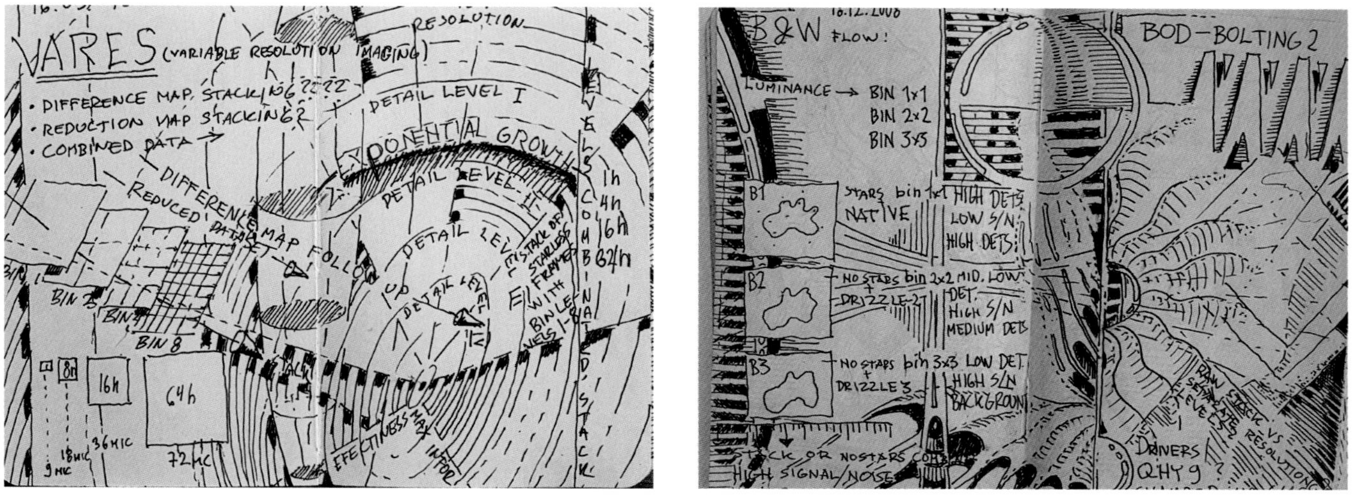

Pages from my notebook, Autumn 2008 – early planning for the project.

I am a perfectionist when it comes to my photography. It is a quality essential for achieving great results, but also a burden. The Milky Way image might have been completed five years earlier if I had been willing to omit some of the extremely faint targets or accept a lower level of detail, but I simply could not do that. When the image was finally finished, I did not dwell on the countless sleepless, freezing nights, but remembered the joy I felt when the most difficult parts were finally complete.

Very soon after I finished this massive imaging project, I began to wonder what it might look like in 3-D. I gathered all the necessary distance and structural data and began the conversion process. The result is a highly realistic 3-D model of the northern Milky Way, with the correct relative distances for various nebulae. It is fascinating to see how, when viewed in three dimensions, objects that appear as pairs in a 2-D image are, in reality, separated by vast distances.

I included accurate distance information for perhaps a thousand or so of the brighter stars. This gave me the idea to draw constellation lines into the stereo image. We can now see the actual three-dimensional shapes of familiar constellations.

The 3-D image also reveals the disc-like structure of the Milky Way itself. Vast formations of dark gas and dust are seen floating between and around nebulae. I was able to place these structures at the correct relative distances. Because they appear silhouetted against bright emission nebulae, they must lie in front of the nebulae – otherwise, they would not be visible.

The difference in distance between the closest object – Sharpless 2-216, a planetary nebula in Auriga located around 420 light-years away – and perhaps the most distant object, the supernova remnant Sharpless 2-224, also in Auriga, at approximately 14,700 light-years, illustrates the immense scale of this 3-D field.

The Milky Way is an endless source of celestial wonders, both scientifically and artistically. As a visual artist, I find this region of the night sky profoundly inspiring. Its intricate structures and mesmerizing forms never fail to captivate me. I could happily spend a lifetime capturing images of this cosmic treasure trove.

J.-P. Metsävainio

The mosaic now seen with the constellations marked on it to show the scale of the image. Remember that the stars that make up the constellations are all nearer to us than the nebulae seen in the stereoscopic photograph.

Lower part of the Milky Way panorama. Note the dark lanes of dust and gas floating between and around the emission nebulae stretching from Cygnus (at bottom) to Cepheus.

Close-up of the upper part of the Milky Way panorama. The closest object visible here is the planetary nebula Sharpless 2-216, located 420 light-years away. The most distant is the supernova remnant Sharpless 2-224, at a distance of 14,700 light-years. The millions of stars in this image are so small at this scale that they appear as a mist of light.

Metsävainio (above) with a large print of the Moon, and (below) with his telescope

The Vera C. Rubin Observatory's view of the Virgo Cluster, part of which formed the front cover of this book.

M51/NGC 5194 seen through the eye of the James Webb Space Telescope.

Index

The Vera C. Rubin Observatory

The Vera C. Rubin Observatory, Chile, under southern skies with the Milky Way arching overhead and the Large Magellanic Cloud, the second nearest galaxy to our own, on the left.

Credits

(t) top; (m) middle; (b) bottom

Cover: (front) RubinObs/NOIRLab/SLAC/NSF/DOE/AURA; (rear) Processing: J.-P. Metsävainio/data acquisition: David Lane; (author portraits) Brian May

p. 2–3 NASA/ESA
p. 5 ESO/B. Tafreshi
p. 7 James Symonds
p. 8 Imaged by David Lane (St Mary's University, Halifax, Nova Scotia, Canada); processed by J.-P. Metsävainio
p. 9 P. Cinzano, F. Falchi (University of Padova), C. D. Elvidge (NOAA National Geophysical Data Center, Boulder)
p. 10 Public domain
p. 11 (t) Public domain (b) Photograph courtesy of the Trustees of Birr Castle Collections
p. 12 King's College London
p. 14 (t) Bureau de Longitudes, Paris (b) Public domain
p. 15 (t) Herschel Museum Bath (b) ESA
p. 16 (t) Johan Hagemeyer (b) AIP Emilio Segrè Visual Archives, Shapley Collection
p. 17 (t) AIP Emilio Segrè Visual Archives (b) AIP Emilio Segrè Visual Archives
p. 18 (t) AIP Emilio Segrè Visual Archives, John Irwin Slide Collection (m) University of Hawaii Institute for Astronomy (b) Brian May
p. 19 (t) Brian May (b) Estate of Sir Patrick Moore
p. 21 NASA, ESA, CSA, STScI
p. 22 Mark Norris, Alston Observatory, University of Lancashire; image and processing J.-P. Metsävainio
p. 25 AIP Emilio Segrè Visual Archives, Physics Today Collection
p. 27 ESA/NASA
p. 29 ESA
p. 31 (t) Trinity College Library, Cambridge University, courtesy of AIP Emilio Segrè Visual Archives (b) William Montgomerie EAO/JCMT
p. 32 (t) NASA/ESA (b) NASA/Bill Ingalls
p. 33 NASA/ESA/CSA/STScI
p. 37 (t) Lucent Technologies' Bell Laboratories, courtesy of AIP Emilio Segrè Visual Archives, Physics Today Collection (b) ESA and the Planck Collaboration
p. 39 W. M. Keck Observatory/Adam Makarenko
p. 40 Event Horizon Telescope Collaboration
p. 42 NASA/ESA/HST/STSci/AUR
p. 44 ESO/P. Horálek (Institute of Physics in Opava)
p. 46 J.-P. Metsävainio (original image by Trevor Dobson)
p. 48 Watkins et al. 2017
p. 49 NASA, ESA, CSA, STScI, Klaus Pontoppidan.
p. 50–59 J.-P. Metsävainio
p. 60 Adapted from Wegg & Gerhard (2013), MNRAS vol. 435, p. 1874
p. 61 Adapted from Kruijssen et al., 2015, MNRAS, Volume 447, p. 1059
p. 65 (t) Ken Crawford
p. 66 (t) EHT Collaboration (b) Copyright Austrian National Library, Vienna/Österreichische Nationalbibliothek Bildarchiv
p. 68–69 Jason Chu
p. 70 NASA/ESA/STSci/Alysson Pagan
p. 73 (t) J.-P. Metsävainio/Mark Norris, Alston observatory, University of Lancashire (b) Mark Norris, Alston observatory, University of Lancashire
p. 75 ESA/NASA/J. Lee and the PHANGS HST Team
p. 76 NASA/ESA/A. van der Hoevan
p. 77 J.-P. Metsävainio
p. 78 Mark Norris/Alston Observatory, University of Lancashire
p. 79 J.-P. Metsävainio
p. 80 NASA/ESA/HST
p. 81 (t) Alston Observatory, University of Lancashire, Mark Norris (b) J.-P. Metsävainio
p. 82–84 J.-P. Metsävainio
p. 84 (t) J.-P. Metsävainio/ Mark Norris/Alston Observatory, University of Lancashire (b) NASA/ESA/Hubble Heritage Team (STSci/AURA)
p. 85 (t) J.-P. Metsävainio (b) NASA, ESA and A. Zezas (Harvard-Smithsonian Centre for Astrophysics); GALEX data: NASA, JPL-Caltech, GALEX Team, J. Huchra (Harvard-Smithsonian Centre for Astrophysics); Spitzer data: NASA/JPL/Caltech/Harvard-Smithsonian Centre for Astrophysics
p. 86 NASA/ESA
p. 88–89 Processing: J.-P. Metsävainio/data acquisition: David Lane
p. 90 Jason Kirk, University of Lancashire, Astrophysical Journal, vol. 798, p. 58
p. 91 (b) J. C. Muñoz/ESO
p. 92 (t) Zdeněk Bardon/ESO (r) ESO/VISTA VMC (b) ESO/IDA/Danish 1.5 m/R. Gendler, U. G. Jørgensen, J. Skottfelt, K. Harpsø (r) Digitized Sky Survey 2
p. 93 ESA/NASA/HST
p. 94 (t) ESO Acknowledgement: VST/Omegacam Local Group Survey (b) J.-P. Metsävainio
p. 95 Fabrizio Francione/Public domain
p. 96 J.-P. Metsävainio
p. 97 (t) J.-P. Metsävainio (b) Rogelio Bernal Andreo
p. 98 ESO
p. 100 James Symonds
p. 102–105 J.-P. Metsävainio
p. 104 Mark Norris/Alston Observatory, University of Lancashire/ J.-P. Metsävainio
p. 105 NASA/ESA Hubble/J.-P. Metsävainio
p. 107 Mike Selby/Mark Hanson
p. 108 J.-P. Metsävainio/Adam Block Mount Lemmon SkyCenter, University of Arizona
p. 109 NASA/ESA/HST
p. 110 NASA/Chandra/J.-P. Metsävainio
p. 111 ESO
p. 112–114 J.-P. Metsävainio
p. 116 J.-P. Metsävainio/CHART 32 team
p. 117 CHART 32 team
p. 118 NASA/ESA and the Hubble Heritage Team (STSci/AURA)
p. 120 J.-P. Metsävainio/NASA
p. 121 J.-P. Metsävainio/Copyright Martin Pugh, Heaven's Mirror Observatory
p. 122 Martin Pugh, Heaven's Mirror Observatory
p. 123 J.-P. Metsävainio/Copyright R. Jay GaBany Blackbird Observatory
p. 129 J.-P. Metsävainio/International Gemini Observatory/NOIRLab/NSF/AURA
p. 130–132 J.-P. Metsävainio ESA/Hubble & NASA
p. 133 (t) Atlas 3D Collaboration (b) Keck Telescopes
p. 135 J.-P. Metsävainio, International Gemini Observatory/NOIRLab/NSF/AURA/L. Bass
p. 136 (t) International Gemini Observatory/NOIRLab/NSF/AURA/L. Bassino; image processing J. Miller (Gemini Observatory/NSF NOIRLab), M. Rodriguez (Gemini Observatory/NSF NOIRLab), & M. Zamani (NSF NOIRLab) (bl) The Observatories of the Carnegie Institution of Washington, courtesy of AIP Emilio Segrè Visual Archives, Tenn Collection (br) SDSS/David Kirkby
p. 138 (t)(m)(b) J.-P. Metsävainio (X-ray NASA/CXC/CfAR/Kraft tl; Sub-mm MPIfR/ESO/APEX/A.Weiss et al.)
p. 139 NASA/ESA/Hubble/reprocessing Raul Villaverde
p. 140 NASA/JPL-Caltech/STSScI
p. 141 AIP Emilio Segrè Visual Archives
p. 142 J.-P. Metsävainio (NASA/ESA/Hubble)
p. 143 NASA, ESA, and the Hubble Heritage (STScI/AURA)-ESA/Hubble Collaboration. Acknowledgment: R. O'Connell (University of Virginia) and the WFC3 Scientific Oversight Committee
p. 144 NASA/ESA/CSA/JWST/SDSS
p. 146 (t) J.-P. Metsävainio/SDSS
p. 149 (t) NASA/ESA/STSci/AURA (b) Rogelio Bernal Andreo DeepSkyColors
p. 150 J.-P. Metsävainio/Adam Block, Mount Lemmon SkyCenter, U Arizona
p. 151 J.-P. Metsävainio (t) Adam Block, Mount Lemmon SkyCenter, U Arizona (b) ESO's New Technology Telescope (NTT)
p. 153 (b) CTIO/NOIRLab/DOE/NSF/AURA
p. 154 (t) J.-P. Metsävainio (b) CTIO/NOIRLab/DOE/NSF/AURA
p. 155 (t) NASA/Hubble (b) SDSS 2
p. 156 (t) A. Fraknoi, courtesy AIP Emilio Segrè Visual Archives (b) NASA/ESA/Hubble, HST Frontier Fields
p. 157 NASA/ESA/Hubble, HST Frontier Fields
p. 158 ESO/Andrew Z. Colvin/N. Bartmann
p. 159 ESO/Andrew Z. Colvin/N. Bartmann
p. 160 Institute of Astronomy, University of Cambridge – Amanda Smith
p. 161 (b) ESA/Euclid/Euclid Consortium/NASA, image processing by J.-C. Cuillandre (CEA Paris-Saclay), G. Anselmi
p. 162 J.-P. Metsävainio/Adam Block/Mount Lemmon SkyCenter/ University of Arizona
p. 164 J.-P. Metsävainio/Volker Springel (Max Planck Institute for Astrophysics) et al.
p. 167 (t) Jeremiah Horrocks Institute, University of Lancashire
p. 169 NASA/ESA/CSA/STSci
p. 170–180 J.-P. Metsävainio
p. 181 (t) J.-P. Metsävainio (b) Studio Heikkala
p.182 RubinObs/NOIRLab/SLAC/NSF/DOE/AURA/P. Horálek (Institute of Physics in Opava)
p. 184 NASA/ESA/CSA/James Webb Space Telescope
p. 190 NSF–DOE Vera C. Rubin Observatory
p. 192 Brian May

It’s a wrap!

From left to right: authors J.-P. Metsävainio, Derek Ward-Thompson and Brian May, designer and illustrator Jamie Symonds, publisher Robin Rees.

Stereoscopic Viewing

If you're familiar with the technique of **parallel free-viewing** of 'side-by-side' stereo pairs, you will have no trouble seeing all the 3-D images in this book. But to get the best immersive effect, use the OWL viewer supplied here.

This hand-held viewer is designed to give you high quality 3-D viewing of the stereo pairs in these pages.

Directions

Remove your LITE OWL viewer from the pocket opposite, and hold it with either hand, locating the thumb on one of the grips near the lower corners. **Position the book in good light, with no shadows on the page.** Bring the lenses of the viewer close to your eyes, and position your head squarely in front of the image pair you've chosen to view. Move to a distance of about 5 inches (120 mm) from the page, and relax the eyes. Don't squint or strain – you won't need to – and forget the idea that you're viewing something close-up. Instead, expect to witness a view of a distant scene through a window, or through binoculars.

Gently move back and forth a few millimeters until the image is in focus, and slightly adjust the tilt of your head if necessary to align it with the page. At this point the eyes should settle into a relaxed position as the full beauty of the three-dimensional image appears. If you still see two flat images, don't give up! Just disengage for a few seconds and look around the room, to remind yourself what that feels like, and then try again, remembering to relax and look into the distance. It will be worth it!

For more help, and information on all kinds of stereoscopic matters ancient and modern, please visit our London Stereoscopic Company website at

www.LondonStereo.com

THE
London Stereoscopic Company, LTD.